Phuoc Nguyen

E-Wheel™ - The New Generation of Pedal Electric Cycles (Pedelecs)

An Integrated Electric Wheel based on all-in-one idea

Anchor Academic Publishing

Nguyen, Phuoc: E-Wheel™ - The New Generation of Pedal Electric Cycles (Pedelecs): An Integrated Electric Wheel based on all-in-one idea, Hamburg, Anchor Academic Publishing 2015

Buch-ISBN: 978-3-95489-457-4
PDF-eBook-ISBN: 978-3-95489-957-9
Druck/Herstellung: Anchor Academic Publishing, Hamburg, 2015

Bibliografische Information der Deutschen Nationalbibliothek:
Die Deutsche Nationalbibliothek verzeichnet diese Publikation in der Deutschen Nationalbibliografie; detaillierte bibliografische Daten sind im Internet über http://dnb.d-nb.de abrufbar.

Bibliographical Information of the German National Library:
The German National Library lists this publication in the German National Bibliography. Detailed bibliographic data can be found at: http://dnb.d-nb.de

All rights reserved. This publication may not be reproduced, stored in a retrieval system or transmitted, in any form or by any means, electronic, mechanical, photocopying, recording or otherwise, without the prior permission of the publishers.

Das Werk einschließlich aller seiner Teile ist urheberrechtlich geschützt. Jede Verwertung außerhalb der Grenzen des Urheberrechtsgesetzes ist ohne Zustimmung des Verlages unzulässig und strafbar. Dies gilt insbesondere für Vervielfältigungen, Übersetzungen, Mikroverfilmungen und die Einspeicherung und Bearbeitung in elektronischen Systemen.

Die Wiedergabe von Gebrauchsnamen, Handelsnamen, Warenbezeichnungen usw. in diesem Werk berechtigt auch ohne besondere Kennzeichnung nicht zu der Annahme, dass solche Namen im Sinne der Warenzeichen- und Markenschutz-Gesetzgebung als frei zu betrachten wären und daher von jedermann benutzt werden dürften.

Die Informationen in diesem Werk wurden mit Sorgfalt erarbeitet. Dennoch können Fehler nicht vollständig ausgeschlossen werden und die Diplomica Verlag GmbH, die Autoren oder Übersetzer übernehmen keine juristische Verantwortung oder irgendeine Haftung für evtl. verbliebene fehlerhafte Angaben und deren Folgen.

Alle Rechte vorbehalten

© Anchor Academic Publishing, Imprint der Diplomica Verlag GmbH
Hermannstal 119k, 22119 Hamburg
http://www.diplomica-verlag.de, Hamburg 2015
Printed in Germany

Abstract

Today, millions of people all around the world are suffering from air pollution, traffic congestion, unsafe traffic conditions, noise, etc. while our entire planet is under thread as the result of climate change. So cycling becomes essential part of the solution to these major problems. There are hundred reasons to choose pedelec: little parking space, clean, green, etc. and definitely faster than car in crowded town.

Pedelecs are in this context very important. They produce no emissions, no noise and they use very little energy at very low cost. They cause no "external costs", they allow avoiding congestion and parking problems. They assure mobility of elderly people and those with health problems. Pedelecs have a positive impact on public health in general and therefore reduce medical costs. They also contribute to sustainable tourism.

This application offers an introduction to the new generation of pedal electric cycles (pedelecs) and its potential for society in the design and technology in terms of industrial design and mechanical engineering. E-Wheel™, *a multi-award-winning patented design*, stands for Integrated Electric Wheel, based on **all-in-one** idea. E-Wheel™ is not just a redesigning of common pedelecs, however, E-Wheel™ and the others will be playing on ever more significant role in our everyday mobility with very positive "support effect" for urban transportation.

Detail CAD data and Finite Element Analysis (FEA) model for both electromechanical and structure analysis are presented in this work and those show that the E-Wheel™ will be take advantage of conventional electric bicycles (e-bikes) or common pedelecs.

Besides, the apply-oriented of brushless motor microcontroller design is also presented. The electrical requirements of the controller (voltage, current, frequency) influence the section of components is fully developed and used to illustrate these methods.

Copyright and Trademark

Copyright © 2015 Phuoc Nguyen. All rights reserved.

E-Wheel™ is unregistered trademark owned by Phuoc Nguyen.

All other products, logos, brand names, registered and unregistered trademarks or companies mentioned are property of their respective owners.

Testimonials

The E-Wheel™ received so **many praises** from *all over the world* as well as in *the Milan Design Event*, for example:

"The 'E-Wheel' is a new type of pedal electric cycle that is integrated all in one, and using technology the accessory is remotely controlled from a mobile phone or other console device, the lightweight product is made from eco friendly materials and enables faster transportation around the city streets."

Birgit Lohmann – CEO/Chief Editor at Designboom/LDA Judge

"The high-tech E-Wheel – created by Phuoc Nguyen, a Vietnamese mechanical design engineering student – can be fixed to conventional bike frames to boost pedal power, with wireless technology also enabling remote control usage." [15]

Danielle Demetriou – British journalist

"Really like your bike wheel design. Any chance to purchase one?"

Robert Duenner – Senior Vice President Wealth Advisor, Morgan Stanley

"...This project could be a revolution in alternative drives for E-bikes. I wish you all the best for your future and I hope you will have many excellent ideas like this one..."

Roland Schneider – Mechanical R&D Engineer, Continental Automotive GmbH

"E-Wheel by Phuoc Nguyen is a redesigned pedal electric cycle (pedelec), that offers riders new experiments when pedaling. It can be integrated into conventional or folding bicycles and uses wireless tech for remote control and battery charge." [36]

Nanette Wong – Writer, Design Milk Magazine

"..."

Achievements

At the beginning of the project, E-Wheel™ has won two prestigious prizes, PR package and sponsorship:

October, 2013: 3rd prize of **ASEANpreneurs Autodesk Design Challenge**, organized by **NUS Entrepreneurship Society** and **Autodesk Inc.** with the first concept of E-Wheel™. [16]

January, 2014: one of twelve award winners of worldwide **Lexus Design Award 2014**, organized by **Lexus International**, co-hosted by **Designboom magazine** and **DESIGN ASSOCIATION NPO**, with panel display in the **Lexus Design Amazing 2014** at the **Milan Design Week 2014** from *April 8th to April 13th, 2014.* [14]

July, 2014: Two months license of **JMAG®** software package was sponsored by **JSOL Corporation, Japan** for 3D advanced electromechanical analysis.

September, 2014: E-Wheel™ was chose to PR for **Autodesk Student Experts Marketing Assets – Autodesk Education Expert** program, by **Autodesk Inc.** USA.

Table of Contents

Abstract 5

Copyright and Trademark 6

Testimonials 7

Achievements 8

List of Figures 11

List of Tables 13

Acknowledgments 14

1 Introduction 17
   1.1   What is a Pedelec? 17
   1.2   Electricity Consumed by a Pedelec 19
   1.3   Why E-Wheel™? 20

2 Brushless Direct-Current (BLDC) Motor Design and Optimization 22
   2.1   Design Strategy and Goals 22
   2.2   Construction and Operating Principle 22
   2.3   Motor Constant – Prediction Methods: Apply Lorentz Force and Faraday Law Method 29
   2.4   Resistance, Losses and Power Rating 33
   2.5   Motor Prototyping Methods 38
       2.5.1   CAD Modeling 38
       2.5.2   Fractional Pitch Magnets 40
   2.6   BLDC Model Analysis and Optimization 42
       2.6.1   2D Finite Element Method Magnetics 45
       2.6.2   3D Advanced FEA using JMAG® Designer v13.1 46

3 Mechanical Design and Prototyping 48
   3.1   Design Strategy and Goals 48
   3.2   Design Constraints and Overall Structures 48
       3.2.1   Design Constraints 48
       3.2.2   Overall Structures 51
   3.3   BLDC Motor Skeleton 52

    3.4   Battery Packs 54

    3.5   Hub 57

    3.6   Axle 63

    3.7   Bearing 64

4   BLDC Motor Controller Design 66

    4.1   Design Strategy and Goals 66

    4.2   Controller Design 66

        4.2.1   Three phase inverter MOSFETs 68

        4.2.2   Bus Capacitor 71

        4.2.3   Gate Driver 72

        4.2.4   Motor Controller 74

        4.2.5   Position and speed sensing 75

        4.2.6   Power Supply 77

        4.2.7   Battery Management 77

Conclusions 79

References 80

Appendices 83

List of Figures

Figure 1: General pedelec structure (Figure Courtesy of OXYGEN Bicycles, United Kingdom) 17

Figure 2: E-Wheel™ complete assembly (commercial rendering image) 21

Figure 3: Simplified BLDC Motor Diagram [28] 24

Figure 4: Ideal sinusoidal vs. trapezoidal back EMF waveforms, normalized to RMS=1 25

Figure 5: BLDC Motor Cross-Section (Source: Redrawn from material furnished by MICROCHIP Brushless DC (BLDC) Motor Fundamentals) [27] 26

Figure 6: Magnetic flux generated by the small coil of wire passing though the magnet (from worldwide web) 27

Figure 7: Six-Step Commutation (Figure Courtesy of MICROCHIP – Brushless DC (BLDC) Motor Fundamentals) [27] 28

Figure 8: (a) A single length of wire in uniform magnetic field (b) A more realistic depiction of the same interaction in a slotted motor 30

Figure 9: A coil of wire moving through a magnetic field that flip direction will generate a back EMF 32

Figure 10: An example of Parameters feature on Autodesk Inventor® Professional 39

Figure 11: An example sketch on Autodesk Inventor® Sketch environment with fully parametric. 39

Figure 12: The special BLDC Motor design (view section 3.3) of E-Wheel™ (rendering image) 40

Figure 13: BLDC motor having fractional pitch magnets and fractional slot 41

Figure 14: Stator winding scheme 43

Figure 15: Circuit graph 44

Figure 16: The flux distribution in the slotted of BLDC motor with the test current of 7.31 A 45

Figure 17: Three-phase 3D Finite Element simulation of the BLDC motor partial model – Magnetic flux density 46

Figure 18: Joule losses graph 47

Figure 19: Measure over-locknut with calipers 49

Figure 20: Disc brake fit info (Source: Redrawn from material furnished by SRAM – Avid disc brake fit info) 49

Figure 21: Adjustable-cone freehub 50

Figure 22: Ringlè Freehub 50

Figure 23: Parts identification of a quick-release mechanism 51

Figure 24: 10 mm thru axle (Figure Courtesy of American Classic) 51

Figure 25: An exploded view of the E-Wheel™ with the special BLDC Motor design (rendering image) 52

Figure 26: BLDC Motor skeleton.. 53

Figure 27: Precision pivot lock bolt drawing (Figure Courtesy of MISUMI USA) 54

Figure 28: Lithium-ion battery cells arrangement in the battery pack 55

Figure 29: The design of the hub with 32 spoke holes (iso left view).................................... 58

Figure 30: The cross-sectional of the E-Wheel™ 59

Figure 31: Loading calculation hierarchical relationships 60

Figure 32: Forces applied to the E-Wheel™ 63

Figure 33: Speed and current control loop configurations for a BLDC motor 67

Figure 34: Electrical waveforms in the two phase ON operation and torque ripple 68

Figure 35: A typical set of power MOSFET I-V curve 69

Figure 36: The bus capacitor is placed across the DC lines, adjacent to the inverter MOSFET (Three phase inverter) 70

Figure 37: DRV8301 Simplified Application Schematic 73

Figure 38: Overall block diagram of Hall sensors and BLDC motor microcontroller............ 76

Figure 39: Block diagram of circuitry in a typical Lithium-ion battery pack......................... 78

List of Tables

Table 1:	The different between kind of electric bicycles (Source: PRESTO Promoting Electric Cycling for Everyone as a Daily Transport Mode) [8]	18
Table 2:	BLDC motor design parameters	29
Table 3:	Parameters used in Rough Analysis of BLDC motor	33
Table 4:	Calculation of the Resistance for BLDC Motor	34
Table 5:	Types of Surface Insulation Resistance and Typical Applications (Source: Selection of Electrical Steels for Magnetic Cores – AK Steel) [32]	36
Table 6:	Total heat capacity calculation for this case	37
Table 7:	BLDC Motor Designed Parameters	40
Table 8:	Winding parameters	42
Table 9:	Bill of Materials of BLDC Motor	44
Table 10:	Performance Comparison between the Rough Calculation and 3D Analysis (at 250 rpm)	47
Table 11:	Specifications of Panasonic NCR18650 cells (Source: Panasonic battery cells technical catalogue) [9]	55
Table 12:	Physical properties of comparison Acetal polymer materials [17]	56
Table 13:	Key dimensions of the hub	62
Table 14:	Some specifications for the IR MOSFET IRFS3006PbF	69
Table 15:	Current flow during the PWM on and off times, assuming the power supply provides only a DC average current (high power supply inductance and/or high frequency)	71
Table 16:	Some Specifications of DRV8301	73
Table 17:	TMS320F28027 Piccolo™ microcontroller simplified specifications [24]	75
Table 18:	Estimating Financial Expenditures, Suppliers and Part/Service Lists for Prototyping	83

Acknowledgments

After exceeding two years and a half on this project, the author would like to express sincerest thanks and appreciation to many individuals, companies, etc. who have contributed to this project.

First of all, the author would like to express special appreciation and thanks to **M.Eng Le Khanh Dien** for supervising and providing technical support for this study. Very few advisors would be willing to give as much freedom and trust to explore a new design as he does.

The author would like to thank **JSOL Corporation, Japan** for providing opportunities to do interesting hand-on **JMAG®** software package. In particular, thanks to **Mr. Bui Le Hung** at **New System Vietnam Co. Ltd.**, the BLDC Motor electromechanical analysis would not have been possible without his support. And also, thanks to **JSOL staff** for corrected BLDC Motor partial model.

The author would like to express deep gratitude to the advice and assistance from good friends, colleagues. Among these are **Ho Nhat Hung, Ngo Xuan Nghiem**, **To Dien Son**, **Dang Thi Bich Ngoc, Ly Thanh Long, Le Hoang Phong** from Ho Chi Minh City University of Technology, **Dao Doc Truong, Quach Dinh Bao, Tran Le Ngoc Linh** from Ho Chi Minh City University of Science, **Dinh Nho Ngoc Lam**, FPT-Arena Multimedia, **Mr. Vu Viet Hung** from John von Neumann Institute – VNUHCM, **Ms. Phan Thi Ngoc Mai** from Vital System Technology Pte. Ltd. (Vietnam RO), and my roommate **Mr. Phan Tan Hoa**.

For the proofreading of certain chapters in this study report, the author would like to thank **Nguyen Hoang Tri** from Ho Chi Minh City University of Technology and Education. Thank you for your many long hours of dedicated work.

Appreciation is also expressed to **Mr. Bjørn Wittenberg**, Program Manager, **Autodesk Inc. USA.** for using this design project to PR for **Autodesk Student Experts Marketing Assets – Autodesk Student Expert** program.

The author would like to express appreciation to the representatives from **Polaris Laser Laminations, LLC** (USA); **Solid Concepts Inc.** (USA); **Proto Laminations Inc.** (USA); **Gilbert Curry Industrial Plastics Co. Ltd.** (UK); **Hiep Luc Co. Ltd** (Vietnam)... In particular, thanks to

Mr. Steve Sprague (Sale Manager of **Proto Laminations Inc.**) for his kindly providing and searching for sponsorships.

Thanks are also extended to **Mrs. Birgit Lohmann**, Lexus Design Award Judge/Designboom Magazine CEO/Chief Editor; **Ms. Danielle Demetriou**, British journalist; **Mr. Robert Duenner**, Senior Vice President, Wealth Advisor at Morgan Stanley; **Mr. Roland Schneider**, Mechanical R&D Engineer, Continental Automotive GmbH; **Ms. Nanette Wong**, Design Milk Magazine writer for glowing testimonials, which contributed important improvements.

Gratitude is extended to **Ms. Maki Kurihara** and **Mr. Manabu Kudo** at Lexus Design Award Secretariat Office (Tokyo, Japan), who offered kindly support with highly responsibility regarding to Milan Design Event.

Thanks to the team at **publisher Diplomica Verlag GmbH** for your hard work and support to turning this study report to tremendous community.

Finally, special recognition goes out to my family, for their understanding, support, encouragement and patience during my pursuit of this design study. To my farther, my mother and my younger brother, thank all three of you for your patience and love you more than you will ever know.

1 Introduction

1.1 What is a Pedelec?

A pedelec (*pedal electric cycle – pedelec* or *electrically power assisted bicycle – ePAS*) is a bike with an electric motor, which supplies power assist only when riders pedal. A sensor (rotation/torque sensor and Hall Effect sensors) measures whether you are pedaling, and passes this information to a controller. This sensor ensures that the motor only provides assistance when the rider is pedaling.

Power is delivered by a battery pack, which can be re-charged through a suitable charger from electric network. Batteries are often mounted to the rack or onto the frame, and sometimes they are built into the frame.

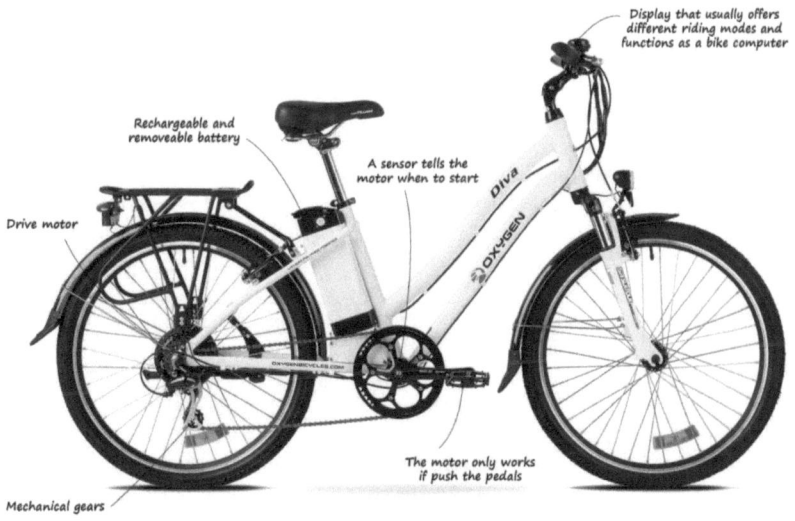

Figure 1: General pedelec structure (Figure Courtesy of OXYGEN Bicycles, United Kingdom)

The German inventor Egon Gelhard invented and patented the pedelec principle in 1982. Unfortunately he could not find a cycle manufacturer willingly implementing his ideas in a product. To be fair, at that time this would have been extremely difficult, because digital motor control and sensor technology were still in the early stages of development, and could not have been manufactured at an acceptable price. So it took another ten years until the Japanese motorbike maker Yamaha developed the first pedelec, and launched it into the Japanese market in 1993. Yamaha understood that with the pedelec they were dealing with a new category of vehicle which only intuitively had anything in common with bikes and motorbikes. [7]

Because of the assistance from motor, riding a pedelec makes users feel that there is wind behind themselves, even when going uphill or in adverse weather conditions. As a result, you never really sweat nor do you get out of breath. Riding a pedelec has a positive influence on your physical condition. You can also choose to ride your pedelec without the help of the electric motor, just as a conventional bicycle.

The advantages go beyond the personal level too, and are relevant to wider society. According to the World Health Organization, 30 minutes of gentle physical exercise are sufficient to extend life by around 8 years*. Pedelec riding can supply this exercise easily. Thus the individual is spared illness, and society is spared costs, through reduced sickness days and increased productivity.

(BICYCLE) PEDELEC	(MOPED) PEDELEC	E-BIKE
Motor only works when pedalling	Motor only works when pedalling	Motor always works
Motor stops at $25\ km/h$ and motor output is $< 250\ W$	Motor works above $25\ km/h$ and motor output is $> 250\ W$	
No further obligation	Age limit, helmet, driving license and insurance obligations depending on your country	Age limit, helmet, driving license and insurance obligations depending on your country
Riders can ride on cycle paths and cycle lanes	In most countries riders are not allowed to ride on cycle paths and cycle lanes	In most countries riders are not allowed to ride on cycle paths and cycle lanes

Table 1: The different between kind of electric bicycles (Source: PRESTO Promoting Electric Cycling for Everyone as a Daily Transport Mode) [8]

Even if no renewable energy is used to charge the pedelec the environmental impact of using a pedelec is probably more positive. This is the case when using a pedelec instead of a

---

* Source: Dr. Günter Klein, *WHO-EHEH* Bonn. *European Centre for the Environment and Health of the WHO.* Presentation: "Wirtschaftliche und menschliche Nutzung körperliche Aktivität im Alltag", 18. 04. 2005. Conference: *Wirtschaft in Bewegung*.

car with a combustion engine fuelled with fossil fuels. The positive impact is due to the high efficiency of the electric motor (80%) in comparison with the low efficiency (25-35%) of the combustion engine.

1.2 Electricity Consumed by a Pedelec

The question now is what influences the annual electricity consumption of a pedelec? The amount of power drawn from the battery during one kilometer of cycling depends on a bunch of factors: [7]

- The chosen *'assistance factor'* which is usually adjustable by the user. The more assistance is requested from the motor the more electricity will be used. Since control systems are programmed to assist the human pedalling the electricity used during a certain time span depends on the intensity of the rider's pedalling. To be exact, it relies on forces which the rider exerts onto the pedals *(ExtraEnergy e. V.)*. In the extreme of no electric assistance, for instance in case of an empty battery, no electricity is used and the pedelec temporarily mutates to a conventional bicycle. The other case is a very high assistance factor reducing the necessity for pedalling to a very low amount. In the extreme of the electric assistance replacing completely the human force pedalling would be no longer necessary (however this would then by definition not be a *pedelec* but an *e-bike* in its 'electricity only mode' or an electric scooter).
- *Slope*: In terms of cycling uphill, the steeper the slope is the more electricity on the average drawn from the battery going increase because the electrical assistance will be higher on the average. Otherwise, someone cycling downhill will need little or even no electric support. In *Engel [2008]* at $15\ km/h$, for a conventional cycle and a flat terrain $60\ W$ are required and, to keep the same speed on a slope of $8°$, $360\ W$ are needed.
- *Speed of the vehicle.*
- *Wind speed and direction:* head wind (opposite the driving direction) increases the aerodynamic resistance, hence leading to a higher electricity demand. In contrast, wind may also drive you from behind (tail wind) and thus reduce the energy consumption.

- *Mechanical and electric efficiencies* of the involved equipment (motor, gear, ergonomic height of the saddle, etc.). Most relevant of all above components are the air pressure in the tires as many cyclists will confirm from their own experience. Low efficiencies lead to an increased electricity demand for the same electro-mechanical assistance.

- The number of *stop-and-go cycles* and of *accelerations*.

Further parameters as specified in *ExtraEnergy e. V.* are the position of the rider during riding and the weight of the ensemble rider and pedelec, e. g. *Engel [2008]* assumes an increase $3\ W$ on power for a rise of $7\ kg$ on weight on a flat terrain at $22\ km/h$ and of $20\ W$ for the same weight difference on a slope of $8°$ inclination at $12.2\ km/h$.

1.3 Why E-Wheel™?

For many, at first glance the pedelec is simply a bicycle with some additional electrics. For the cycle industry, the design opportunities are first limited by the reference to the fundamental "bicycle" concept, and second by the technical options available for production. However, the author feels that the pedelec is much more than that, at least, from the viewpoint of a Designer. It is not only a form of transportation but also it is an expression of the status and lifestyle of their owners. The task of a Designer for a product which already exists is to make them better – more functional, more beautiful and more practical. Therefore, why it is built like a conventional bike – but with a motor? Why we do not shift to a new level!

This is the main inspiration for the first pedelec concept for nearly two years and a half, it is called E-Wheel™ (abbreviation of *Integrated Electric Wheel*) – An integrated electric wheel based on all-in-one idea. With the new cutting edge of design pedal electric cycle (pedelec) – E-Wheel™ will give the rider new experiments when pedalling. The designs of E-Wheel™ are integrated all of the mechanical and electrical components in a hub very elegantly and be fitted as a kit to almost any conceivable bike. The core objective remained unchanged.

E-Wheel™ are suitable for a wide variety of people who are not able or willingly ride a conventional bicycle or who simply need faster transportation and most efficient means of transport in town.

Figure 2: E-Wheel™ complete assembly (commercial rendering image)

E-Wheel™ is not only eco-friendly but also very lightweight. It is integrated **all-in-one**: include 250W (rated power) electric motor, 36V – 8.9Ah – 320Wh Lithium-ion battery, electrics drive. E-Wheel™ can run up to 60km on one fully charged* and the maximum speed is 25 km/h. E-Wheel™ has the wireless technology for remote control and battery charge. You can control E-Wheel™ by your mobile phone (or control console).

Copenhagen Wheel (superpedestrian.com) [10] is the first success design of retro-fitting pedelec, since their initial announcement in 2009. But the drive is not the main focus, rather the gathering of sensor data on air quality and networking to data centers and other users.

With some initial successes, the author would like to inspire the communities to rethink their attitudes to mobility in a visually dazzling way. At this point, some familiar designs were announced, for example: *FlyKly Smart Wheel* [11], *Zehus BIKE+* [12], and *Electron Wheel* [13], etc. The author sure that many talented designers will push it as far as possible. However, the E-Wheel™ still have unique features which are used in a unique way in the context of overall design.

---

* Maximum range based on one battery full charge, using 150% assistance mode, according to use in ideal conditions. Distance will vary depending on road conditions, riding surface, cyclist's weight and required assistance.

2 Brushless Direct-Current (BLDC) Motor Design and Optimization

2.1 Design Strategy and Goals

The ability to design motors to fit specific applications is an opportunity that is, from the point of view of author, highly valuable and yet also not well-known. To most, the process of designing electric motor-based systems involves digging through catalogs of motors, immediately limiting the design space to a set of existing components. By the time this set is filtered by *physical constraints* and *performance requirements*. Breaking down the black-box status of electric motors to open up new design options is the primary goal of this study. There is also a significant learning opportunity in designing a *'fit-for-purpose'* motor, which may have played an even larger role in the author's motivation to pursue such projects.

Useful analysis techniques include combined CAD/FEA using solid modeling and finite element magnetic simulation will be clarified.

In summary, the goals of this design study are to:

- Evaluate the conditions under which a custom motor design may be called for.
- Demystify the design of custom brushless motors by showing simple to advance analysis and simulation techniques as applied to the case studies.
- Provide, though this design, some examples of modern rapid prototyping techniques for making custom motors.

This chapter approaches the simplest methods, only requiring high school physic skills to design the BLDC motor.

2.2 Construction and Operating Principle

BLDC motors are a type of synchronous motor. This mean that the magnetic field generated by the stator and the magnetic field generated by the rotor rotate at the same frequency. Compare with the induction motors, BLDC motors do not "slip".

BLDC motors come in single-phase, 2-phase and 3-phase configurations. Corresponding to its type, the stator has the same number of windings. This application only focuses on 3-phase outer motors.

Stator:

The stator of a BLDC motor consists of stacked steel laminations with windings placed in the axial slots distribute along the inner periphery. Most BLDC motors have three stator windings connected in star configuration. Each of these windings are constructed with numerous coils interconnected to form a winding. One or more coils are placed in the slots and they are interconnected to make a winding. Each of these windings are distributed over stator periphery to form an even numbers of poles.

Rotor:

The rotor is made of permanent magnets and can vary from two to many pole pairs with alternate North (N) and South (S) poles. Based on the required magnetic field density in the rotor, the proper magnetic materials are chosen to make the rotor. Ferrite magnets are traditionally used to make permanent magnets. As the technology advances, rare earth alloy magnets are gaining popularity. The ferrite magnets are less expensive but they have the disadvantage of low flux density for a given volume. In contrast, the alloy material has high magnetic density per volume, thus enabling the rotor to compress further for the same torque. Also, these alloy magnets improve the size-to-weight ratio and give higher torque for the same size motor using ferrite magnets.

Neodymium (Nd), Samarium Cobalt (SmCo) and the alloy of Neodymium, Ferrite and Boron (NdFeB) are some examples of rare earth magnets. Continuous research is ongoing to improve the flux density to compress the rotor further.

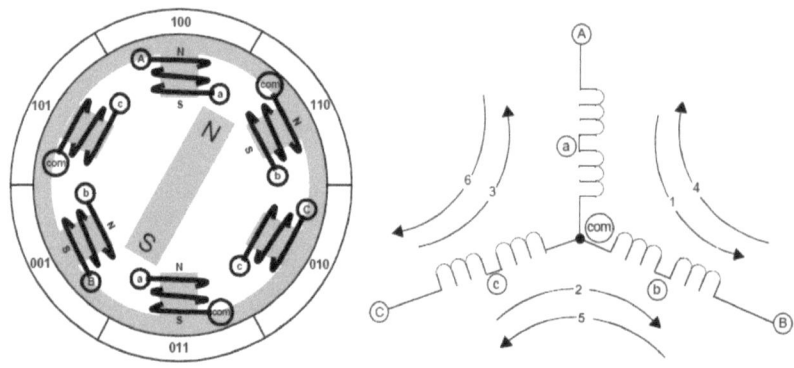

Figure 3: Simplified BLDC Motor Diagram [28]

What is back EMF?

When a BLDC motor rotates, each winding generates a voltage known as back Electromotive Force (or back EMF), which opposes the main voltage supplied to the windings according to Lenz's Law. The polarity of this back EMF is an opposite direction of the energized voltage. Back EMF depends mainly on three factors:

- Angular velocity of the rotor
- Magnetic field generated by rotor magnets
- The number of turns in the stator windings

$$back\ EMF\ (E) = NlrB\omega$$

Where N is the number of winding turns per phase, l is the length of the rotor, r is the internal radius of the rotor, B is the rotor magnetic field density and ω is the motor's angular velocity.

Once the motor is designed, the rotor magnetic field and the number of turns in the stator windings remain constant. The only factor that governs back EMF is the angular velocity or speed of the rotor and as the speed increases, back EMF also increases.

Trapezoidal or Sinusoidal?

The differentiation is made on the basis of the interconnection of coils in the stator windings to give the different types of **back EMF**.

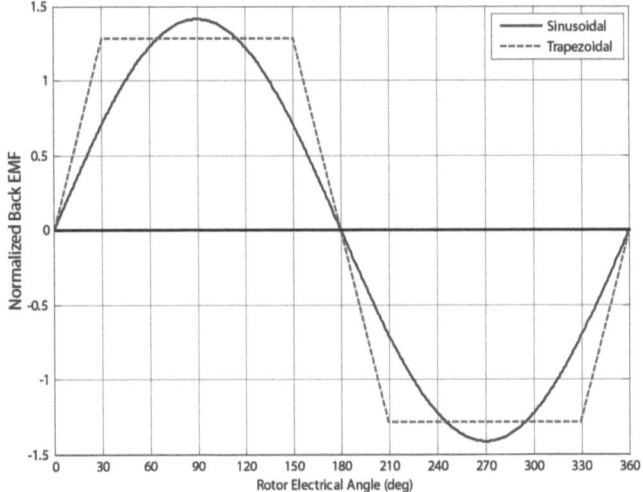

Figure 4: Ideal sinusoidal vs. trapezoidal back EMF waveforms, normalized to RMS=1.

As their names indicate, the trapezoidal motor gives a back EMF in trapezoidal form and the sinusoidal motor's back EMF in sinusoidal. In addition to the back EMF, the phase current also has trapezoidal and sinusoidal variations in the respective types of motor. This make the torque output by a sinusoidal smoother than that of a trapezoidal motor. However, this come with an extra cost, as the sinusoidal motors take extra winding interconnections because of the coils distribution on the stator periphery, thereby increasing the copper intake by the stator windings.

Some physical conditions that would lead to a trapezoidal back EMF are:

- Concentrated windings.
- No stator or magnet skew.
- Discrete magnet poles with uniform magnetization.

These conditions all lead to sharp transitions in the flux linkage, which give the trapezoidal back EMF waveform its distinct shape. While it does influence the shape of the back EMF, stator core saturation is *not* the reason for the flat top of the back EMF. (Saturation implies small rate of change of flux, so it would affect the shape near the back EMF zero crossing, not at the peaks.)

Hall Effect Sensors – *Why we need Hall effect sensors for this BLDC motor???*

The Hall effect has been known for over one hundred years, but has only been put to noticeable use in the last three decades. Today, Hall effect devices are included in many products, ranging from computers to sewing machines, automobiles to aircraft, and machine tools to medical equipment.

The commutation of BLDC motor is controlled electronically. To rotate the BLDC motor, the stator windings should be energized in a sequence. It is also important to know the rotor position in order to understand which winding will be energized for the following step of energizing sequence. The rotor position is sensed by using Hall effect sensors embedded into the stator. Most BLDC motors have three Hall effect sensors embedded into the stator on the non-driving end of the motor.

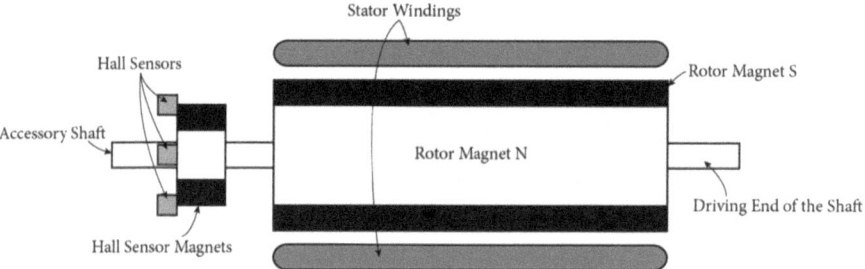

Figure 5: BLDC Motor Cross-Section (Source: Redrawn from material furnished by MICROCHIP Brushless DC (BLDC) Motor Fundamentals) [27]

Whenever the rotor magnetic poles pass near the Hall sensors, the give the high or low signal, indicating the N or S pole is approaching the sensors. Based on the combination of three Hall effect sensor signals, the exact sequence of commutation can be determined.

Theory of Operation:

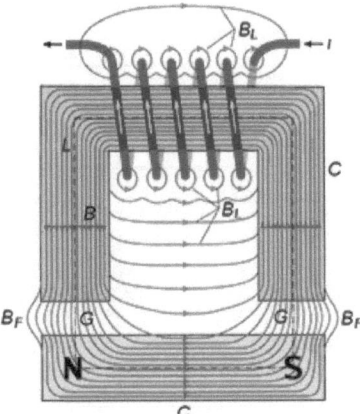

Figure 6: Magnetic flux generated by the small coil of wire passing though the magnet (from worldwide web)

Each commutation sequence has one of the windings energized to positive power (current enters into the winding), the second winding is negative (current exists the winding) and the third is none-energized condition. Torque is produced because of the interaction between the magnetic field generated by the stator coils and the permanent magnets. Ideally, the peak torque occurs when these two field are at 90° to each other and falls of the field move together. In order to keep the motor running, the magnetic field produced by the windings should shift position, as the rotor moves to catch up with the stator field. What is known as *"Six-Step Commutation"* defines the sequence of energizing the windings as bellows:

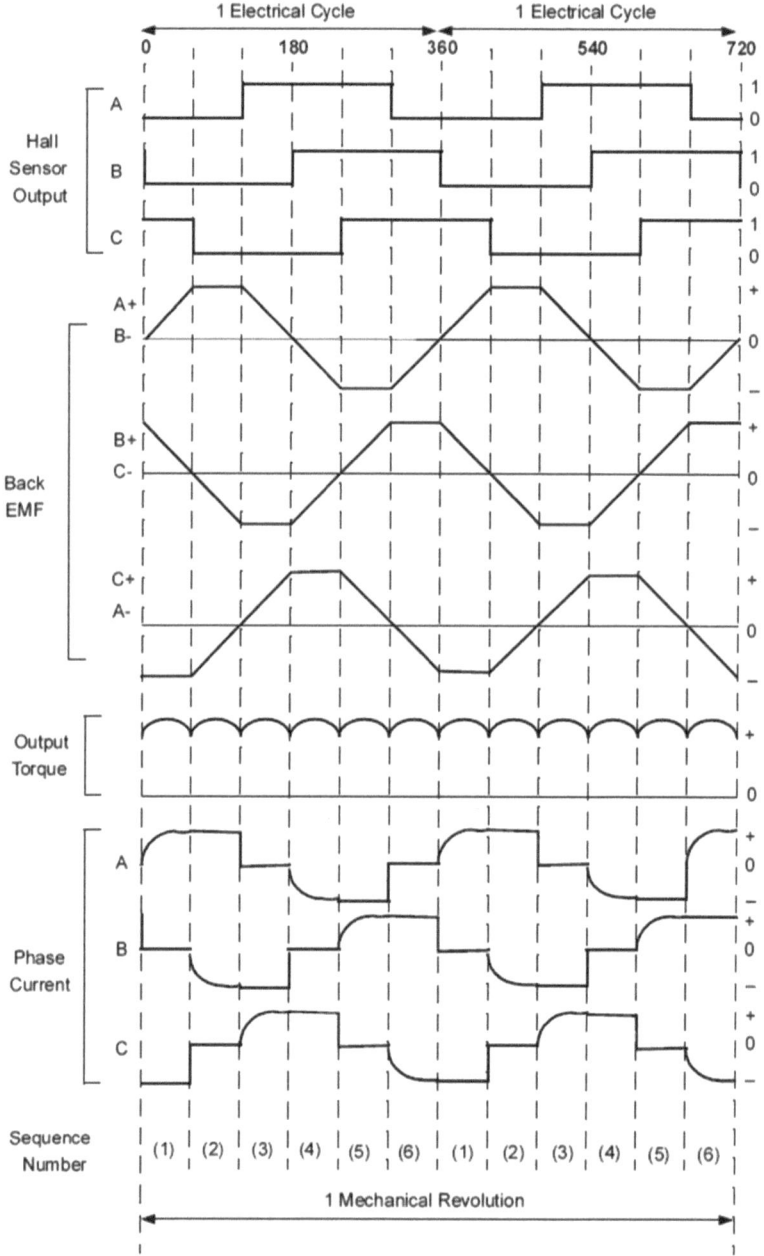

Figure 7: Six-Step Commutation (Figure Courtesy of MICROCHIP – Brushless DC (BLDC) Motor Fundamentals) [27]

2.3 Motor Constant – Prediction Methods: Apply Lorentz Force and Faraday Law Method

Initially, E-Wheel™ are designed and optimized for 26 inches rim, with the top speed of bike is 25 kilometers per hour (6.94 meters per second), and achieved by the designed power 250 Watts. For easy to calculate, the author chose the outer speed is always fixed at 250 rpm.

Top speed (kph)	25
Power supply (V)	36
Rim (inch)	26
Design power (W)	250

Table 2: BLDC motor design parameters

Ideal angular velocity is constant:

$$speed\ at\ motor\ outer = \frac{\pi \omega_{outer} \times D_{outer}}{60} = \frac{\pi \times (250\ rpm) \times (167.1\ mm) \times 10^{-3}}{60}$$
$$= 2.19\ m/s$$

$$Total\ force\ produced = \frac{designed\ power}{speed\ at\ motor\ outer} = \frac{250\ W}{2.19\ m/s} = 114.29\ N$$

The author used the combination of two methods to rough estimate the motor constant in static DC analysis using the Lorentz force and Faraday's Law on a wire. There are two simplest method, requiring only high-school physics, and work well for this case.

Firstly, apply **Lorentz force formula**:

$$\vec{F} = I\vec{L} \times \vec{B}$$

Where I is the current passing through the wire, L is a vector representing the length of wire and direction of positive current flow, and B is the strength and direction of the magnetic field. The force acts in a direction that is perpendicular to both the vector of the field and the current, following the right-hand rule.

The below figure show how the theoretical Lorentz law scenario can be extended to a real, slotted motor. In the slotted motor, turns of wire are wound around a steel core and do not directly interact with the magnetic field in the air gap. However, since the effect of the steel core is to channel flux through the winding, it is possible to define a fictitious current loop that flows around the edge of the steel *in the air gap*. If there are N turns of wire carrying current I, this fictitious current carries NI. The fictitious current NI interacts with the air gap magnetic field according to the Lorentz force law.

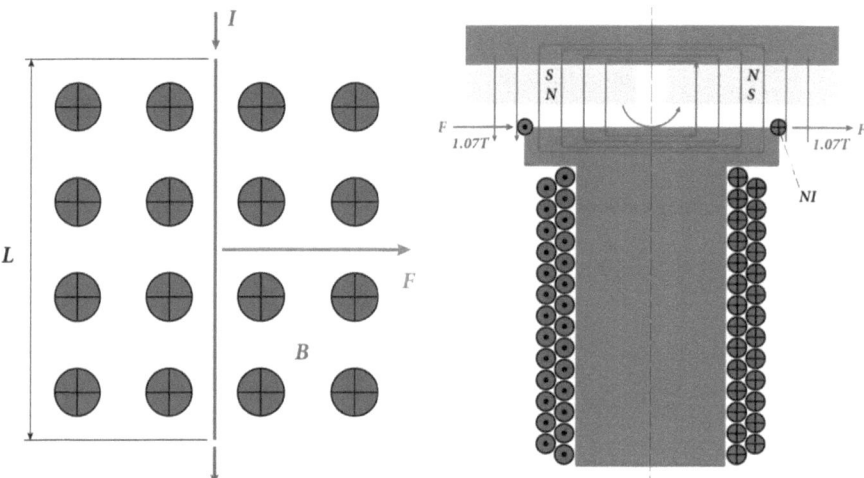

Figure 8: (a) A single length of wire in uniform magnetic field (b) A more realistic depiction of the same interaction in a slotted motor

There are several important consequences of this extension of the Lorentz force law to the slotted motor. For one, the electromechanical conversion occurs at the air gap. Also, the force acts on the steel stator core, not the windings themselves. The windings are shielded from the magnetic field.

$$F = 2NILB$$

The value N is the total number of *active turns* in the motor. The factor of 2 accounts for both halves of a coil. The author assume the motor is driven using six-step commutation, then two of three phases are active at any given time. Thus, N would be two times the number of turns per phase.

The length, L, is called the *active length*, for a slotted motor it is the length of the stator lamination stack. (It does not include the length of end-turns, which don't contribute to the magnetic interaction.)

The field strength, B, is the parameter with the most uncertainly. Typically, the only starting point for estimating this would be the remnant flux density rating of the magnets. For NdFeB magnets, this value B_r, can be on order of $1.2 \div 1.4T$. For this application, the magnets used were grade N42 (NMX-41SH – Hitachi Metals) [2], which have a B_r of approximately $1.3T$. The actual air gap flux will be significantly less than this. One way to estimate peak air gap flux is with a ratio of magnet thickness to total air gap:

$$B_{peak} = B_r \frac{t}{t+g} = 1.07\ T$$

Where t is the magnet thickness and g is the air gap dimension. (For example, if the air gap were equal to one magnet thickness, the peak air gap field would be roughly half the remnant field strength.)

The current passing through the wire would be equal 7.31 Amps.

$$F = (current\ though\ the\ wire, I) \times (active\ turns, N) \times (active\ length, L) \\ \times (magnetic\ field, B)$$

$$\Rightarrow active\ turns, N = \frac{(114.29\ N)}{(7.31\ A) \times 2(25 \times 10^{-3}\ m) \times (1.07T)} = 293$$

So, the number of windings per tooth is 10 turns.

Secondly, there is another way to estimate the motor constant using the **Faraday's Law**, which describes the back EMF generated due to changing flux linkage by a coil of wire and the back EMF can be calculated:

$$Back\ EMF, E\ (or\ V) = \frac{d\phi}{dt} = B\frac{d}{dt}(A_{north} - A_{south}) = \Delta BLv$$

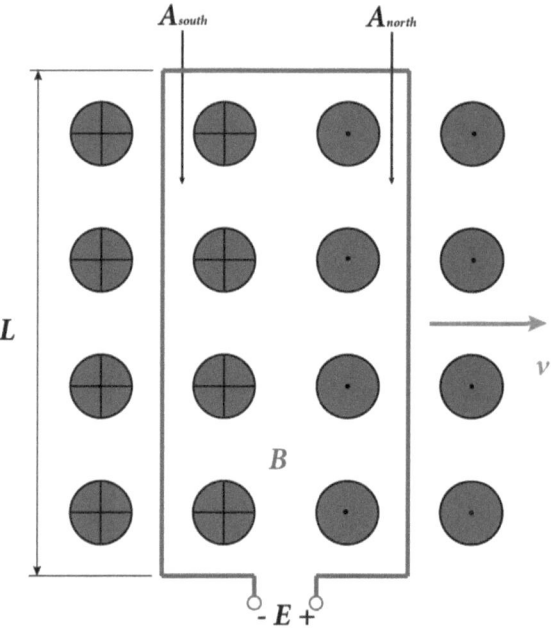

Figure 9: A coil of wire moving through a magnetic field that flip direction will generate a back EMF

In this equation, v is the linear velocity of the wire coil.

$$Induced\ voltage, V = (2 \times 1.07\ T) \times (25 \times 10^{-3}\ m) \times (293\ turn) \times (2.19\ m/s) = 34.2\ V$$

Thus, the motor constant is:

$$K_t = \frac{V}{\omega_{no-load}} = 1.626\ \left(\frac{V}{rad/s}, \frac{Nm}{A}\right)$$

The BLDC motor of the E-Wheel™ used wye-connected windings, which are easy to analyze in this rough estimate since it is possible to pass current through only two phases.

The below table shows a list of parameters necessary to carry out the rough analysis of BLDC motor, as well as the resulting K_t estimate.

Parameters	Value
N (active turns)	293
B_r (magnetic remnant flux density)	1.3 T
t (magnet thickness)	3 mm
g (air gap dimension)	0.65 mm
$B_{peak} = B_r \times t/(t+g)$	1.07 T
B (field used for Lorentz force law)	1.07 T
L (active length)	25 mm
$K_t = 2NLB$ (for wye)	1.626 Nm/A

Table 3: Parameters used in Rough Analysis of BLDC motor

2.4 Resistance, Losses and Power Rating

One of the sources of power loss is the resistance of the motor winding. The resistance dissipates energy while the inductance stores energy and back EMF converts energy.

The total area of the winding is the number of turns times the cross-sectional area of a single wire. It is some fraction of the total area occupied by the windings, not including space between them. Increasing this fraction, called the "fill factor", is a major goal in motor design. Hand-winding, however, limits the design to a relatively low fill factor, usually much less than 50%.

The winding resistance is straightforward to predict and measure. By defining different input parameters, we could calculate the optimal winding pretty exactly. To get the maximum efficiency, we should try to store as much as copper as possible in the stator slots and keep the electrical resistance of the winding as low as possible – which means that we should use wire with the largest possible gauge. Yet, this is very hard to wind without damaging the stator plates and isolation coating in the wire.

The windings don't really convert power with 100% efficiency. They have some resistance that generates heat. (Similarly, the steel will also not conduct the magnetic field perfectly.) To get a rough estimate of the loss in the copper, the author used the rating for 20AWG magnet wire resistance, which is about $33.31\ m\Omega/m$, over the length of two phases of windings, which is approximate 15 meters. And the heat generated by the copper losses is:

$$P_{cop} = I^2 R = (7.31\ A)^2 (0.5\ \Omega) = 27\ W$$

Compare this to a 25W light bulb. It is a lot of heat, including other losses, such as in the steel and some friction in the bearings, but the motor could still be over 80% efficient when running at 7.31 A!

The continuous power rating is of course related to the motor's ability to reject heat to the outside environment, which depends on many factors and would require a much more thorough thermal study to fully characterize. For now, the heuristic is simply applied to this motor design to get ballpark continuous power rating. *Table 4* shows how this simple analysis goes for the design.

Parameters	Unit	Value
Turns per Phase (Series/Parallel)	s/1p	140
Wire Gauge	AWG	20
Wire Diameter	mm	0.812
Total Parallel Cross-Section Area	mm^2	0.812
Resistance per Unit Length	mOhm/m	33.31
Wire Type		Round
Winding Termination		Wye
Average Length per Turn	mm	53.2
Total Length per Phase	mm	7448
Estimated Resistance (wye)	mOhm	248.09
Ampacity by 9Arms/mm^2 (wye)	Arms	7.31
Operating Voltage	V	11.40
Heat Generated by 2phase $\times$ I^2 $\times$ R	W	27
Total Power OUT	W	223
Percent Power Loss to 2phase $\times$ I^2 $\times$ R	%	10.61

Table 4: Calculation of the Resistance for BLDC Motor

Other losses are mechanical P_{mech} in nature. Coulomb friction aside, most sources of drag and damping increase with the speed of the motor. Viscous drag in the bearings is one example and aerodynamic drag in the air gap is another.

Core loss P_{core} is the electrical power expended in the form of heat within the core of electrical equipment when those cores are subjected to alternating magnetizing force. This, of course, is incidental to the production of the desired magnetic flux. According to classic magnetic theory, core loss is considered to be composed of several types of loss. These are hysteresis loss, eddy current loss within individual laminations, and inter-laminar losses that may arise if laminations are not sufficiently insulated from one another.

The power dissipated by eddy current is also inversely proportional to the resistance of the conductor experience the eddy current. Thus, eddy current loss in silicon steel is generally lower than in copper, due to its higher electrical resistance.

The use of thin laminations (and/or small wire gauge) is primarily to minimize eddy current. By decreasing the size of conductive "loop", the voltage and current induced in the conductor can be minimized. Power dissipation scales as the square of conductor thickness. The BLDC motor, 0.47 millimeters (0.0185 inches) non-oriented silicon steel laminations (M-19 AISI Grade) with C-5 insulation coating were used in the stator core. The term "non-oriented" means the magnetic properties are practically the same in any direction of magnetization in the plane of the material.

Despite the difficulties in precisely determining required surface insulation resistance, the problem cannot be resolved by using more insulation to be sure resistance is adequate. The below table indicates the types of insulations that generally prove to be adequate for a variety of typical applications. These insulations will normally provide adequate resistance assuming normal or average fabricating conditions.

ASTM Insulation Designation	Description	Typical Applications
C-0	The natural oxide surface of flat-rolled silicon steel which gives a slight but effective insulating layer sufficient for most small cores and will withstand normal stress-relief anneal of finished cores by controlling the atmosphere to be more or less oxidizing to the surface. Available on all nonoriented grades.	Fractional horsepower motors, pole pieces and relays, small communication power transformers, and reactors.
C-2	An inorganic insulation consisting of a glass-like film forms during high-temperature hydrogen anneal of grain-oriented silicon steel as the result of the reaction of an applied coating of MgO and silicates in the surface of the steel. This insulation is intended for air-cooled or oil-immersed cores. It will withstand stress-relief annealing temperatures and has sufficient interlamination resistance for wound cores of narrow width strip such as used in distribution transformer cores. It is not intended for stamped laminations because of the abrasive nature of the coating. Available on oriented grades only.	Wound- core, power frequency devices, distribution transformers, saturable reactors and large magnetic amplifiers.
C-3 (AK Steel Core Plate No. 3)	An enamel or varnish coating with excellent insulation resistance intended for air-cooled or oil-immersed cores. The C-3 coating will enhance punchability, is resistant to normal operating temperatures, but will not withstand stress-relief annealing. Available on fully processed nonoriented grades only.	Air-cooled, medium-sized power and distribution transformers; medium-sized continuous-duty, high efficiency rotating machinery. Can be oil cooled.
C-4 (AK Steel Core Plate No. 4)	A chemically treated or phosphated surface intended for air-cooled or oil-immersed cores requiring moderate levels of insulation resistance. It will withstand stressrelief annealing and serve to promote punchability. Available on fully processed nonoriented grades only.	Applications requiring insulation similar to C-3 and a stress-relief anneal. Used extensively for small stamped laminations requiring higher resistance than provided by annealing oxides.
C-5 (For Fully Processed Nonoriented Electrical Steels)	An inorganic/organic insulation. It will withstand stress-relief annealing in a neutral or slightly reducing atmosphere. Available on fully processed cold rolled nonoriented grades.	Principally intended for air-cooled or oil-immersed cores which utilize sheared laminations and operate at high volts per turn. Also finds applications in apparatus requiring high levels of interlaminar resistance.

Table 5: Types of Surface Insulation Resistance and Typical Applications[*] (Source: Selection of Electrical Steels for Magnetic Cores – AK Steel) [32]

---

[*] Based on the classification of surface insulations by ASTM (A 976).

Although C-5 insulation coating can slightly reduce atmosphere affected, this does not mean that the laminations steel core can face up with moisture issues for areas of high humidity such as in tropical environments, typically rainy Asia Pacific. The motor is not watertight, this breathe though the hub cover. The motor becomes warm due to operation which causes pressure to build up inside as the volume of air expands and vents out. The opposite is true when we finish riding; the motor cools, it creates a vacuum and sucks air back in. The moisture is trapped inside and collects. The renewed heat of operation creates steamy Amazon-like rainforest conditions and will corrode all unprotected surfaces, including the magnets, the stator core.

E-Wheel™ is the enclosure-system, which means the phase wires is very short, thus, cable losses P_{cable} could be neglect. The total efficiency is calculated as below:

$$\eta = \frac{P_{design} - (P_{mech} + P_{cop} + P_{cable} + P_{core})}{P_{design}}$$

Thermal is the most important in re-arrange the battery cells and choose the best material for battery packs and also optimize the outer motor. Heat generated by the motor will be absorbed by the good contact parts, especially is the aluminum hub shell (outside) and the battery packs (inside) before being transferred out to the environment. In any case, a **temperature rise of 100°C** is allowed, by far than normal operating conditions.

The total heat capacity of the motor is calculated from a sum of heat capacities for the windings and any part of the motor in good thermal contact with the windings. Thus, in the E-Wheel™, the steel lamination core is a larger factor in heat capacity. The below table summarizes the calculation of total heat capacity:

	Mass (kg)	Heat Capacity, cp(J/kgK)
Copper Mass	0.168	384
Steel Mass	0.605	460
Aluminium Mass	0.425	910
Total Heat Capacity C, Ws/K		729.516

Table 6: Total heat capacity calculation for this case

The time for a 100°C temperature rise is:

$$t = \frac{C \times 100K}{3I^2R}$$

where I is the RMS (Root Mean Square) current, R is phase resistance, and C is the total heat capacity shown in the above table.

2.5 Motor Prototyping Methods

Advances in computing, modeling, and display tools have increased the speed and accuracy with which communication, visualization, and analytical problems are performed. More complex designs can be produced more quickly with better functionality and fewer errors with the **three-dimensional (3D) modeling**.

In this section, the focus have on how to use CAD software to modeling BLDC motor for analysis and simulation in the next phase of design progress.

2.5.1 CAD Modeling

The design cycle of a prototype motor build can be significantly reduced by integrating computational resources. Deeply understanding of these progresses are the best way to save time and money. Especially, with the short cycle techniques that can be used in labs (or by individuals) for which motor design is the primary focus. In fact, software packages of Autodesk Inc. played an important role in this project. The E-Wheel™ was designed in the **Autodesk Inventor® Professional** (Student version) CAD software with all dimensions driven by equations. With over hundred parameters, this seem very complex, very hard to organize. However, that means the author could play with the model geometry, with the benefit of making small or big changes to analyze the impact on the performance.

Figure 10: An example of Parameters feature on Autodesk Inventor® Professional

Using **3D modeling** (solid modeling), the calculation of important mechanical properties of parts and assemblies can be done easily. The volume that a part or assembly occupies usually can be calculated with a single command after the computer model has been built and assigned materials. Properties of volume, such as mass, center of gravity, moments of inertia, and principal axes, can also be calculated. Without a solid modeler, the calculation of these properties would be laborious, especially for complex geometries.

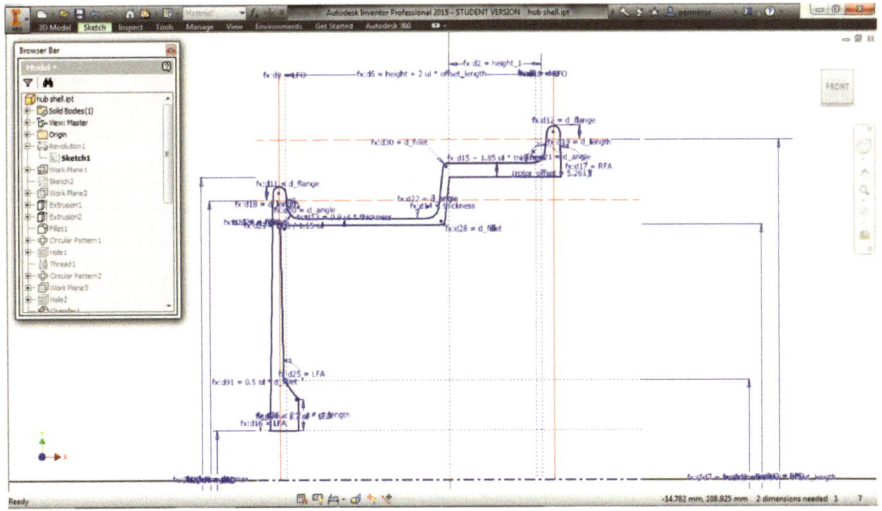

Figure 11: An example sketch on Autodesk Inventor® Sketch environment with fully parametric.

39

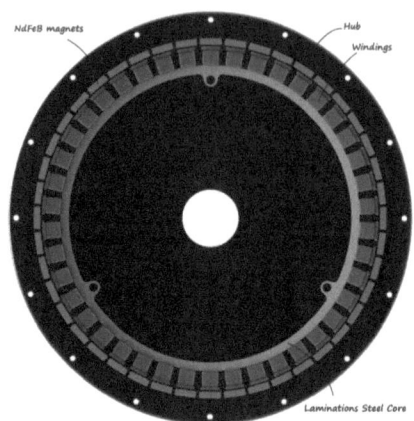

Figure 12: The special BLDC Motor design (view section 3.3) of E-Wheel™ (rendering image)

In sum, Autodesk Inventor® Professional seems to be an incredibly useful tool that the author would have used more extensively had time and manpower allowed it.

Three-dimensional modeling software is also used to build geometric models that can be exported for finite element analysis (FEA) (in the next section). FEA is a numerical analysis method used to calculate results such as stress distribution, temperature distribution, or deformation in a part.

All	Outer Diameter (mm)	167.1
	Gap Length (mm)	0.65
	Height *(mm)*	25
Outer Rotor	Number of Poles	44
	Outside Diameter (mm)	167.1
	Inside Diameter (mm)	156.3
	Magnet width (mm)	10
	Magnet Thickness (mm)	3
	Magnet Corner R (mm)	0.5
Inner Stator	Number of Slots	42
	Outside Diameter (mm)	155
	Inside Diameter (mm)	138
	Tooth Width (mm)	4.5
	Tooth Fang Width (mm)	8.6
	Tooth Fang Thickness (mm)	1.5
	Core Back With (mm)	3.9265
	Teeth Top Radius R (mm)	0
	Teeth Fillet (mm)	0.5

Table 7: BLDC Motor Designed Parameters

2.5.2 Fractional Pitch Magnets

To understand the impact of fractional pitch magnets, here, there are gaps θ_t between magnet poles containing nonmagnetic material. Therefore, the angular magnet pitch θ_m is smaller than the angular pole pitch θ_p. [5]

The BLDC motor have discrete magnet poles with uniform magnetization, this condition leads to sharp transitions in the flux linkage, which give the trapezoidal back EMF waveform its distinct shape. In general, low-cost motors tend to have more characteristics that cause to a trapezoidal back EMF.

For any given rotor position, compute the flux linkage to the stator coil. By identifying key positions where flux transitions occur, the flux linkage can be plotted. Differentiating the flux linkage gives the back EMF, which has the same shape as the torque produced by the coil under constant current conditions.

Figure 13: BLDC motor having fractional pitch magnets and fractional slot

The rotor is the hub and bonded to the inside are a series of magnets with alternating poles facing inward and outward. Design with as much of magnet poles is equivalent to achieve more torque and low speed.

2.6 BLDC Model Analysis and Optimization

To understand deeper the analysis and estimate of the motor constant, we must know about exact shape of the back EMF, which depends on the flux distribution. The analyses have also many factors influence the shape of the back EMF, such as:

- Magnet geometry, air gap dimension.

- Winding factor. The distribution of winding is as important as that of the magnets, since it determines how much flux gets linked.

- Steel lamination stack geometry and properties, such as saturation and hysteresis.

- Steady/Transient thermal: very important in re-arrange the lithium ion battery cells and choose the best material for battery packs.

Finite element method magnetic solvers can include all of these effects, producing a more accurate estimate of the flux distribution in the motor. This can be used to establish a better estimate of the motor constant, or to verify the rough calculations.

For this case study, a finite element analysis (FEA) was conducted using FEMM 4.2 [1], and JMAG® Designer v13.1 [2] for 3D advanced analysis. The FEA was tightly integrated with the mechanical computer aided design (CAD) of the motors themselves. With the motor geometry and materials in place, a test current can be passed into the windings and the resulting torque can be computed by the FEA solver. This method using current and torque to establish K_t.

Connection Type	Star Connection
Series Number	14
Parallel Number	1
Number of Turns	9
Setting Type	Phase Resistance
Winding	Auto Winding
Number of Layers	2
Coil Pitch	1

Table 8: Winding parameters

The BLDC Motor can be wound using Distributed LRK* (DLRK) with the following diagram (42 slots and 44 magnet poles) *(Table 7)*: **AaAaAaABbBbBbBCcCcCcCAaAaAaABbBbBbBCcCcCcC** [31] where:

- "A" and "a" are first phase wire U
- "B" and "b" are second phase wire V
- "C" and "c" are third phase wire W
- Capital (upper case) letter means Clockwise
- Small (lower case) letter means Counter-Clockwise

Three phases are connected by Star (wye) connection. Connect together: Start A - End C, Start B - End A , Start C - End B. In wye connection system gives more torque and uses fewer amps, system 1.73 less turns needs to be wound to get the same power as Delta system does.

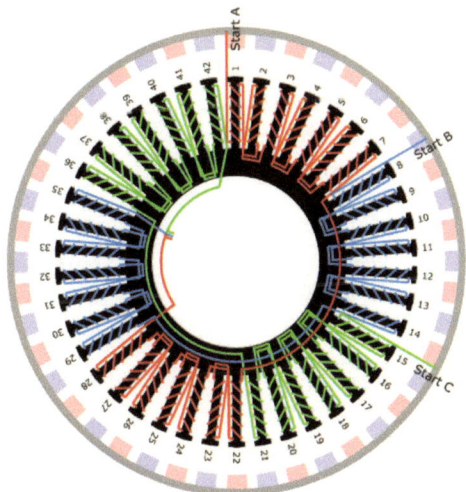

Figure 14: Stator winding scheme

---

* The LRK motor was developed by three gentlemen named Lucas, Retzback and Kuhfuss. In a nutshell, the LRK winding diagram is about trying to obtain the highest flux (maximum lines of magnetic force) with a given amount of stator metal and magnetism.

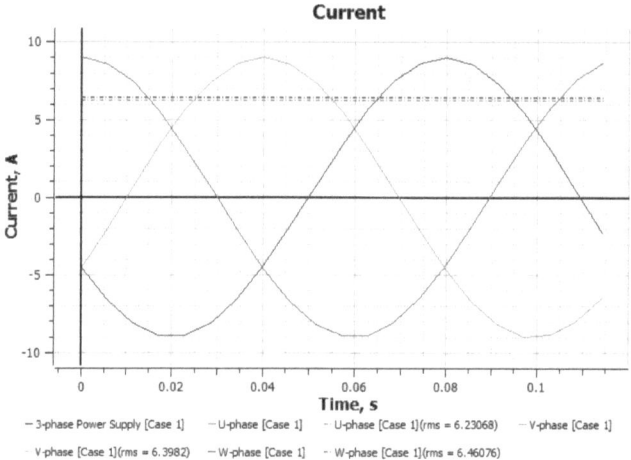

Figure 15: Circuit graph

To FEA demonstrating, we must define exact materials for each part of BLDC motor as be shown in the below table:

Rotor Core	Category	JSOL-Conductor
	Type	Aluminium
	Density	2.699
Rotor Magnet	Category	Hitachi Metals-NdFeB Magnet
	Type	NMX-41SH
	Temperature	60
	Temperature Correction Factor	0
	Magnetization Pattern	parallel
	Flux Density (T)	1.3
Coil	Category	Copper
	Density	89.6
Stator Core	Category	JSOL-Steel sheets
	Type	50A230
	Density	7.6
Common Material Properties	Iron Loss Correction Factor	1

Table 9: Bill of Materials of BLDC Motor

This BOM is based on the available materials with the highest characteristics and competitive price.

2.6.1 2D Finite Element Method Magnetics

Simple 2D analysis is enough in the spirit of low cost prototyping like a free 2D finite element package, FEMM 4.2. With the motor geometry and the materials in place, the solver can produces a map of the flux density.

The figure below shows the FEA result of BLDC motor on FEMM with the test current of 7.31 A. Viewing the flux distribution and reveals that the rotor (outer) and the stator (inner) are integrated to the magnetic circuit, carrying flux between magnets in some comparison models:

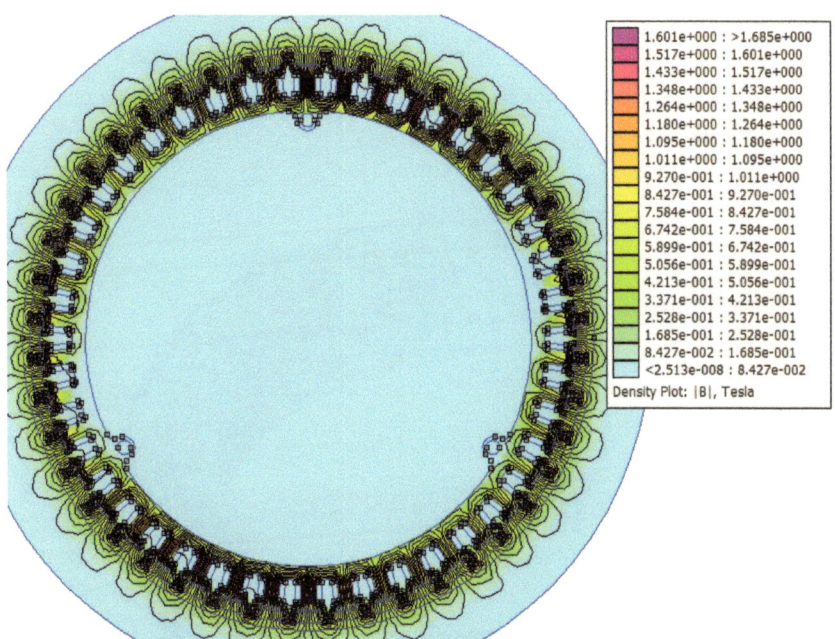

Figure 16: The flux distribution in the slotted of BLDC motor with the test current of 7.31 A

Unlike the rough analysis (using formula calculations), the FEA can capture the torque variance as a function of rotor angle. For now, the simplest thing to do is to assume the motor will be driven by six-step commutation *(Figure 7)*, so that at any time there is one optimum commutation state that produces the most torque. With FEA, it is quick to find this state by testing all six cases. Within this 60° sextant, the torque will still vary somewhat (as in real life), but at least it will be a good estimate.

2.6.2 3D Advanced FEA using JMAG® Designer v13.1

3D finite element analysis model is take advantage of 2D model. 2D analyses may be adequate concerning modeling geometry and the main magnetic flux in a motor but some phenomena (thermal, leakage, skew, etc.) are inherently 3D effects, thus requiring 3D analyses. However, setting up a 3D model is more complicated than 2D model.

For example, the below figure shows how a solid model of a BLDC motor are created for 3D analysis. The same model is then used to generate a FEA mesh in preparation for an analysis of the magnetic flux density distribution in the structure. The flux densities are calculated and their contours are plotted directly atop the original solid model image to show the location and magnitude of the flux densities in the motor. *Figure 17* shows the flux density distribution of the BLDC motor at a specific time step.

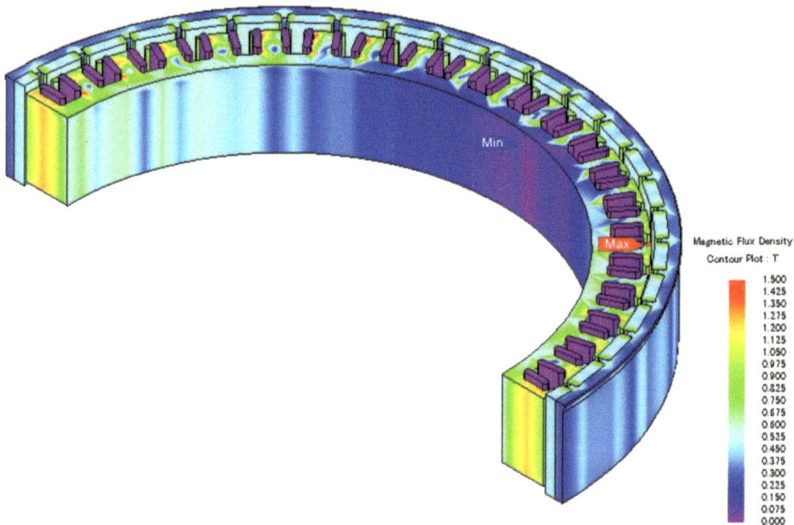

Figure 17: Three-phase 3D Finite Element simulation of the BLDC motor partial model – Magnetic flux density

After 'trial-and-error' analyzing the model many times, the author limited the RMS current to the maximum of 9 A/mm^2 to increase efficiency of the BLDC motor and the below figure shows the Joule losses of the copper coils. [3]

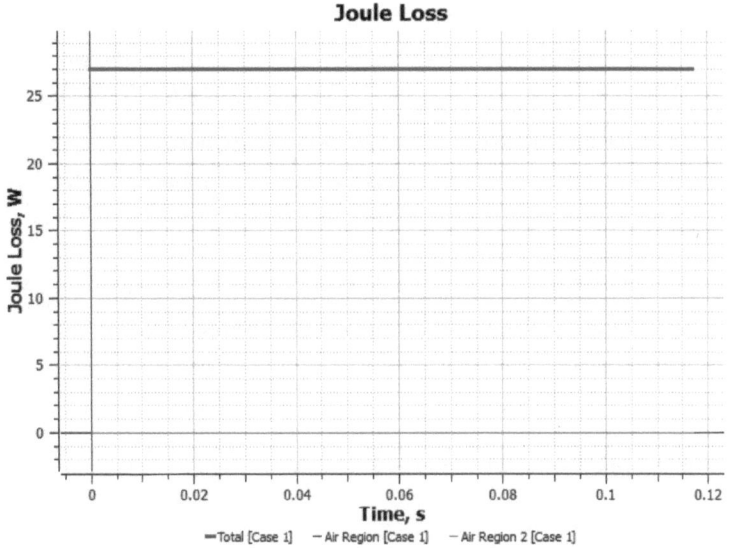

Figure 18: Joule losses graph

The below table is a short comparison between mathematical formula calculations and FEM analysis result from JMAG® Designer software. [6]

	Rough Calculation	3D Analysis	Delta (%)
Torque (Nm)	10.0	10.0	0.0
Mechanical power (W)	250.0	250.0	0.0
Joule losses (W)	27.0	27.0	0.0
Efficiency (%)	89.2	89.2	0.0

Table 10: Performance Comparison between the Rough Calculation and 3D Analysis (at 250 rpm)

This design seems to be a high efficiency that need not much further evolution or more improvement than fully experiments from the prototype.

3 Mechanical Design and Prototyping

3.1 Design Strategy and Goals

The design of a whole system requires attention to the design and selection of individual components (shaft, bearing, etc.). However, as it is often the case in design, these components are not independent. They are closely relate to each other. This design based on parts/spare parts which available in production and easy to find out in stores. Lightweight and conventional structure remains unchanged are strongly highlight in the mechanical design stages.

Moreover, the compliance with previous standards as well as unique features make E-Wheel™ more acceptable. The key of this design is the harmony combined of standardization and individualization.

3.2 Design Constraints and Overall Structures

With the term *all-in-one* keep in mind, E-Wheel™ was designed for retro-fitting pedelec and could be integrated all of the electrical components very elegantly and be fitted as a kit to almost any conceivable bike. So, it was subject to the following constraints:

3.2.1 Design Constraints

Hub over-locknut dimension:

Providing more space for a greater number of sprockets increases the offset of the right flange to the left, which in turn significantly increases the wheel's vulnerability (weak and easily hurt) to failure when exposed to high lateral loads (generally only experienced during crashes or other forms of losing control of the bike). In some cases, this is compensated for by adding space to the left side of the hub. A standard seven-speed hub might have 130mm overall spacing, but be more available in a 135mm for both road and MTB bikes. The design of E-Wheel™ based on the most popular over-locknut dimension (OLD) of 135mm, thus, this is the main of overall design constrains.

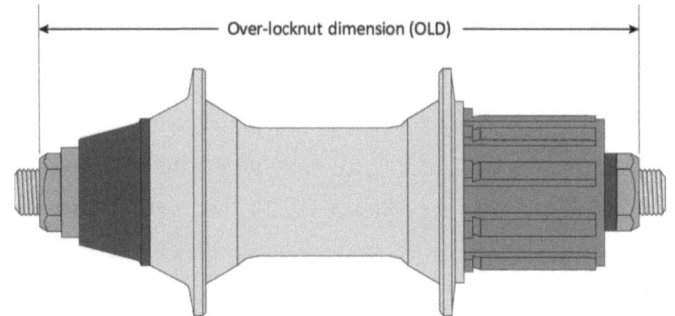

Figure 19: Measure over-locknut with calipers

Disc brake fit:

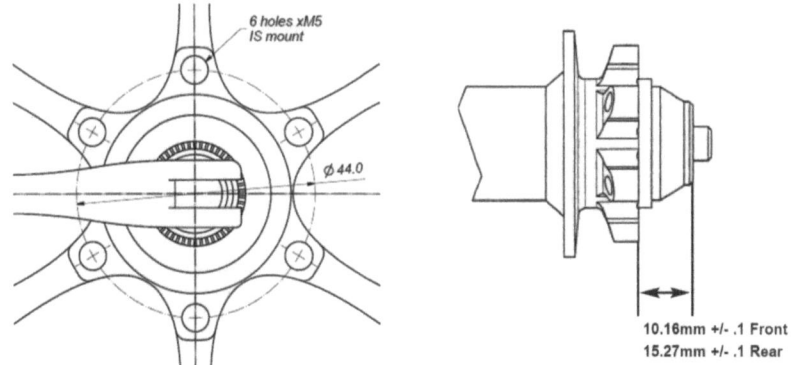

(a) Hubs need to have the International Standard 6 bolt, 44mm bolt circle mounting flange.

(b) The hub flange off-set for the front is 10.16mm and the rear is 15.27mm

Figure 20: Disc brake fit info (Source: Redrawn from material furnished by SRAM – Avid disc brake fit info)

Assembly & Disassembly:

Normally, bicycles are used adjustable-cone hubs. Adjustable-cone hubs have a threaded axle, loose balls or balls in a retainer, cones threaded onto the axle, and cups fixed inside the hub shell *(Figure 21)*. This includes adjustable-cone front hubs, adjustable-cone rear hubs that accept a thread-on freewheel, and freehubs (rear hubs that have the freewheel integrated into the hub) *(Figure 22)*. [33]

Disassembling the first end of the axle is easier if the axle is not free to turn. The ideal way to do this is to hold the end of the axle, which is not being disassembled, in a bench vise.

However, when securing the axle in a vise, it is easy to damage either the axle or the locknut. It could not be a daily routine.

For these reasons, in this design, E-Wheel™ was used cartridge-bearing hub. The designs of cartridge-bearing hub is various, with almost every manufacturer designing hubs in very different way, but they all used the cartridge bearing that is pressed into the hub shell. Hadley and Conrad are names that are sometimes used for the cartridge bearing.

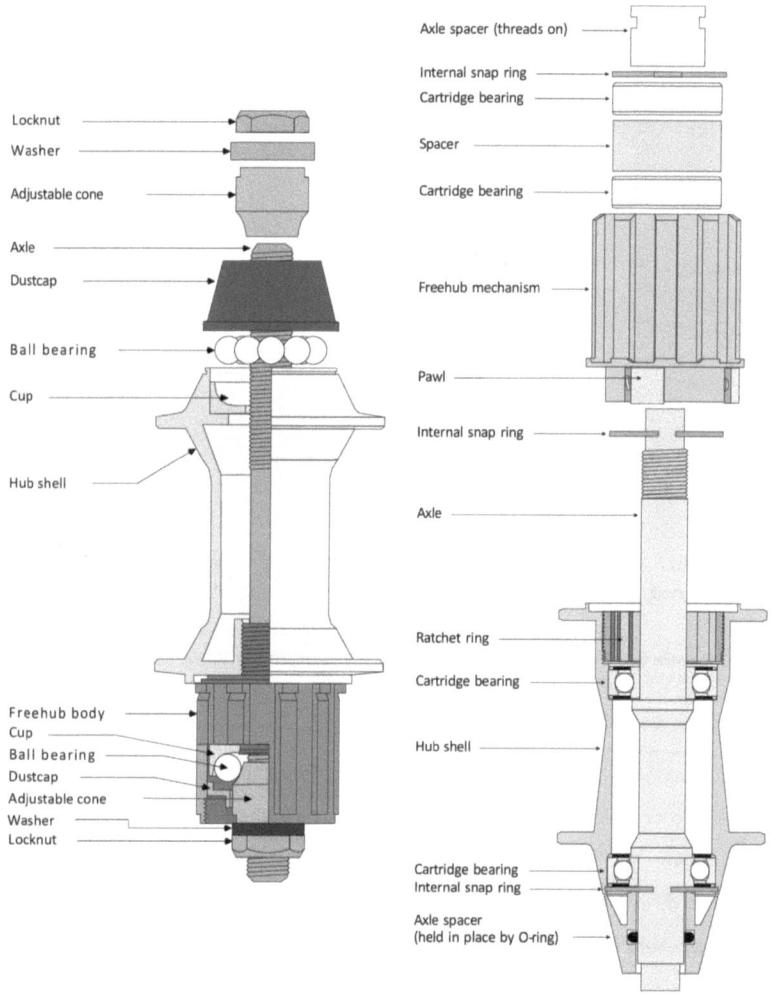

Figure 21: Adjustable-cone freehub   Figure 22: Ringlè Freehub

Using quick-release mechanism takes advantage of solid axles and nuts, the wheel can be easily and quickly removed from the bicycle frame and replaced without using tools by opening and closing cam lever. Another useful way is using thru-axles, it is more stiffer and safer than normal quick-release. Thru-axles are quite accepted in the mountain bike (MTB).

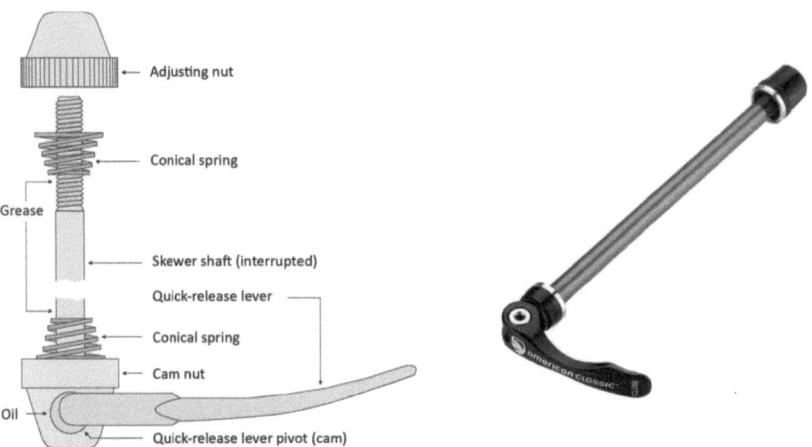

Figure 23: Parts identification of a quick-release mechanism

Figure 24: 10 mm thru axle (Figure Courtesy of American Classic)

3.2.2 Overall Structures

Simplicity is the best place to start a design. A key to system simplicity is the use of modules. A module may be designed with a rigid frame to enable it to combining together and also support machine rigidity. In addition, the packaging of individual elements into modules can have a great effect on overall system performance.

Consideration should be given to the method of assembling the components onto the shaft, and the shaft assembly into the frame (or hub in this case). When components are to be press-fit to the shaft, the shaft should be designed so that it is not necessary to press the component down along length of shaft. Consideration should also be given to the necessary disassembly the components from the shaft. The press fit assembly also allows all hubs to work in almost any currently existing axle configuration, making them compatible with almost every bike.

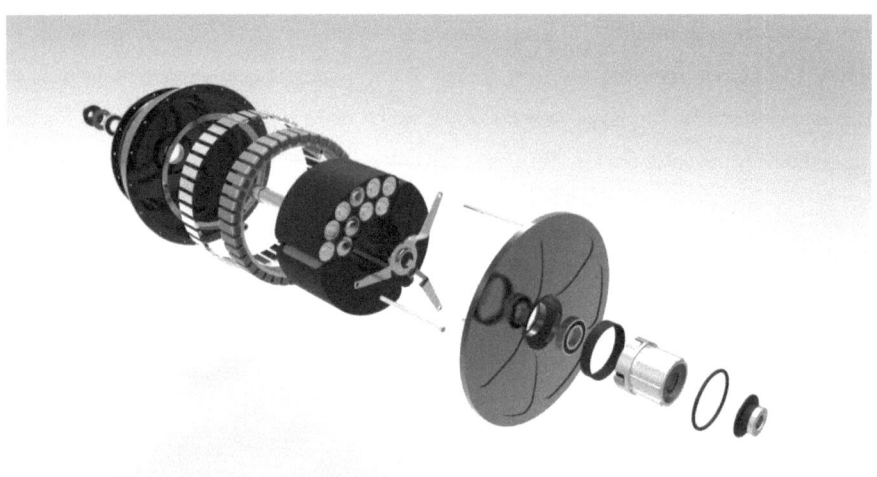

Figure 25: An exploded view of the E-Wheel™ with the special BLDC Motor design (rendering image)

E-Wheel™ was designed as an enclosed-system including 250W (rated power) BLDC motor, 36V Lithium-ion battery, multi-sensor (Hall effect sensors, rotation sensor, temperature sensor), electric drive for retro-fitting as a kit to almost any conceivable bike. The E-Wheel™ is compatible with rear cassettes: single-speed, 5-speed or up to 6-speed via Shimano Uniglide® spline freehub. The batteries pack is designed as separated modules for stability, durability and fitted inside the BLDC motor. This is a very clever design because it completely eliminates the need for wires to be run along the frame. In other words, no more wires are used!

3.3 BLDC Motor Skeleton

The special BLDC Motor structures was inspired incidentally, when the author studied the magnetic flux density of regular electric motors. I realized, they did not across the central of stator steel core, and the highest flux density is distributed outer, at winding tooth only, in which the interaction between permanent magnets and the energized-winding occurred.

That is why the E-Wheel™ is featured with special BLDC Motor structures, i.e. not only for lightest weight but also for the reduction of cost and the simplify of design. However, the stability and durability are still maintained. In addition, I began with the simple case of heat transfer from this operated motor to the surrounding environments. Energy transfer is generally proportional to the surface area, that is, the heat transfer rate may be increased by increasing the surface area across which convection, radiation and conduction occurs. The special BLDC motor with large surface area (in good contact) absolutely take advantage of normal motors structure. Controlling the heat transfer is one key to fabricate materials with enhanced properties.

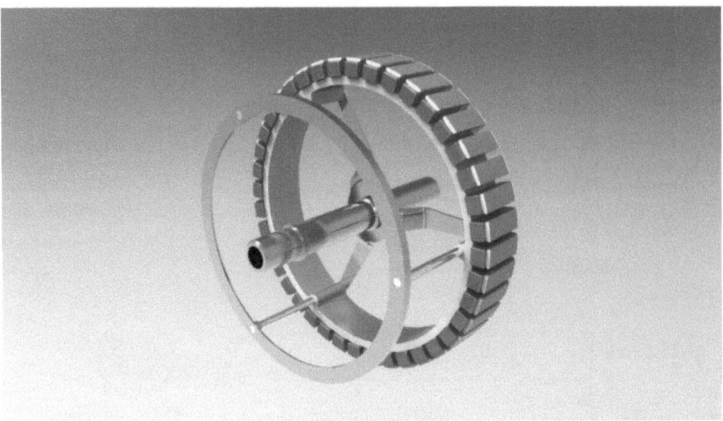

Figure 26: BLDC Motor skeleton

A structure is only as good as the materials and manufacturing processes used. Different materials often motivate the use of different types of manufacturing methods, which affects the design of the structure. For example, if the functional requirements of the structure include high stiffness, so deformations will not affect component alignment and function, a high modulus of elasticity material cannot be desirable. A metal shape with a large cross-sectional moment of inertia might be best. However, an additional functional requirement, such as to provide a large planer surface on which components can be mounted, could make creating the structure just from metal very challenging if the weight were to kept reasonable.

In fact, the stack length of the rotor and the stator are depend on length of NdFeB magnets. The length of 1 inch (approximate 25 millimeters) is the most popular [26]. These magnets are glued inside the rotor hub using epoxy adhesives. No magnet poles of the same kind are next to each other (the polarity changing from **N** to **S** with every position). Using the popular sizes of rare earth magnets is one key to achieve low-cost and easy to do in-house prototyping. The length of stator core is 22.09 millimeters (included 47 laminated steel with 0.47 millimeters (0.0185 inches) thick, non-oriented silicon steel laminations (M-19 AISI Grade) and 2 high temperature isolation plates (inserts for each side of the stator) laminations steel are stacked together by laser welding (or TIG welding). Unlike traditional forms of welding, the laser produces a highly concentrated heat source which enables precise, narrow welds. It is a versatile process which can be used to weld stainless steel, carbon steel, or aluminum. Laser welding is normally used for smaller sized cores like this case. The stator segment lamination is a high-cost, high-risk item of the prototype and require the power of the original design.

The stator is fixed by 3 lock bolts (Precision Pivot Pins) which have diameter of 4mm custom-made and are supplied by **MISUMI** (USA), they have extra low hex socket head and made by 1045 carbon steel as the drawing below:

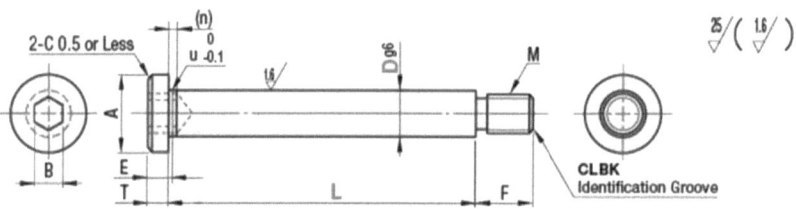

Figure 27: Precision pivot lock bolt drawing (Figure Courtesy of MISUMI USA)

The placement of the pin holes in the stator core at first seemed arbitrary; However, viewing the flux distribution *(Figure 14)* plotted by FEMM (Finite Element Method Magnetic) reveals where the flux density is lowest, they will interfere least with the magnetic circuit. Otherwise, by placing 3-bolt pattern as a self-locking system minimal radial dimensions.

3.4 Battery Packs

In the past few years, Lithium Iron Phosphate (LiFePO4) was the most preferable for transportation vehicles. This day, Lithium-ion battery cells have become essential as a power source for cordless equipment, such as smartphones, laptops, that supports a ubiquitous society. Although Lithium-ion battery is not the most ideal, but up to 95% of the material they contain can be recycled in existing facilities and the storage capability is higher [7]. On the other hand, the continuous development in battery technology are coming out of laboratory has made Lithium-ion higher energy density, more safety and more long life. The most common all round cell type is the

18650 design, which are organized in 3-block, 10-cell per block, 120° per sector, mounted extremely special and useful packed inside the stator and not for removable as shown in *Figure 28*.

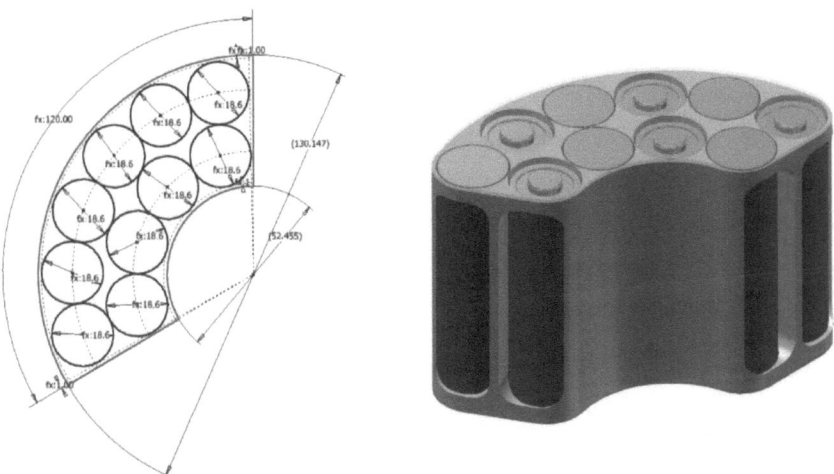

Figure 28: Lithium-ion battery cells arrangement in the battery pack

The arrangements of the battery cells like a puzzle: *"How to arrange as much as circles onto the fixed frame with the specific sizes?"* This sounds very interesting and of course a little bit hard!

The design of E-Wheel™ based on Panasonic Lithium-ion NCR18650 battery cells with the specifications as the following:

Nominal Voltage		$3.6\ V$
Nominal Capacity*	Minimum	$1900\ mAh$
	Typical	$2000\ mAh$
Dimensions	Diameter	$Max\ 18.6\ mm$
	Height	$Max\ 50.0\ mm$
Approximate Weight		$33.5\ g$

Table 11: Specifications of Panasonic NCR18650 cells (Source: Panasonic battery cells technical catalogue) [9]

Lithium-ion battery cell is very sensitive. Overcharging, short circuit, physical damage raises the risk of overheating and igniting. Any electric devices (include E-Wheel™) require more robust and safer batteries. That is why E-Wheel™ has self-locking battery packs while Lithium-Ion

---

* Charge: constant voltage/constant current, 4.2 V, max. 1330 mA, 38 mA cut-off – Discharge: constant current, 380 mA, 2.5 V cut-off – Temperature: 25 °C

battery manufacturers (*like Panasonic – one of the leading Lithium-Ion battery manufacturers in the world*) have employed **Heat Resistance Layer** (**HRL**) technology to improve the safety of Lithium-ion batteries significantly. This heat resistance layer consists of insulating metal oxide on the surface of the electrodes which avoid overheating if short circuit occurs. [9]

Physical damage can be reduced by using appropriate structure for the battery packs. In this, thermal is the most important criterion for re-arranging the battery cells and choosing the best material for battery packs and also optimizing the outer motor. Heat generated by the motor will be absorbed by the good contact parts, especially the aluminum hub shell (outside) and the battery packs (inside) before being transferred out to the environment. In any cases, a **temperature rise of 100°C** is allowed, by far than normal operating conditions!

Heat causes expansion which would lead to improper constraint, that can further cause an unstable situation that lead to failure of the battery packs. All need here is the material with high allowable service temperature, low coefficient of linear thermal expansion and also directly dissipate the heat transferred to extremely sensitive battery cells inside. [18]

In approaching of the materials selection, the battery packs are made by polyacetal: Acetal Homopolymer/Delrin or Acetal Copolymer supplied by **Quadrant Engineering Plastic Products** (Quadrant EPP). They have high mechanical strength, stiffness and better wear resistance (especially thermal-oxidative degradation).

Properties	Acetal Homopoly-mer/Delrin®	Acetal Copolymer
Density (g/cm^3)	1.43	1.41
Coefficient of linear thermal expansion (average value between 23 and 100°C) $m/(m.K)$	110×10^{-6}	125×10^{-6}
Temperature of deflection under load – method A: 1.8MPa (°C)	115	105
Max. allowable service temperature in air – continuously: 5000/20000h* (°C)	105/90	115/100

Table 12: Physical properties of comparison Acetal polymer materials [17]

---

* Temperature resistance over a period of 5,000/20,000 hours. After these periods of time, there is a decrease in tensile strength of about 50% as compared with the original value. The temperature values given here are thus based on the thermal-oxidative degradation which takes place and causes a reduction in properties. Note, however, that, as for all thermoplastics, the maximum allowable service temperature depends in many cases essentially on the duration and the magnitude of the mechanical stresses to which the material is subjected.

QUADRANT Engineering Plastic Products' stock shapes can be easily machined on ordinary metalworking and in some cases on woodworking machines. In view of the poor thermal conductivity, relatively low softening and melting temperatures of thermoplastics, generated heat must be kept to a minimum and heat build-up in the plastics part avoided. This is in order to prevent deformations, stresses, color changes or even melting. [29]

Therefore:

- tools must be kept sharp and smooth at all times,
- feed rates should be as high as possible,
- tools must have sufficient clearance so that the cutting edge **only** comes in contact with the plastics material,
- a good swarf removal from the tool must be assured,
- coolants should be applied for operations where plenty of heat is generated (e.g. drilling).

Besides the most outstanding design features, the used materials are very important in E-Wheel™. Materials utilization was totally a selection process that involved choosing from a given, rather limited set of materials the one best suited for an application by virtue of its characteristics.

3.5 Hub

The hub may appear to be the most important part of the wheel because it is centrally located and all other wheel components rotate around it. In fact the hub acts only as an anchor for the spokes and is a fairly static part of the structure. Although bearings, axles and freewheels involve many clever design features, the structural parts of the hub that affect the wheel are the flanges. Although flanges appear simple, their design can have important effects on hub function.

Figure 29: The design of the hub with 32 spoke holes (iso left view)

The hub are made by *Aluminum 6061-T6, black oxide* finishing with a high (large) flange on the right and a low (small) flange on the left have been made in an attempt to counteract rim offset in multispeed rear wheels and arrangement has effect to fitted in BLDC motor inside.

The large-flange, however, makes spoke insertion on the low side difficult. High-lows cannot reduce vertical loads, the principal cause of spoke failures. Torque loads have so little effect on fatigue that high-low hubs offer no improvement over conventional hubs. However, in this case, high-lows are the most effective way to achieves lightweight and contributes to an exclusive design.

Flanges must be strong enough to support spoke loads, yet be softer than the spokes. Although steel is stronger than aluminum, it does not support spokes as well because it is too hard. Aluminum alloy has adequate strength, is lighter than steel, and is soft enough to allow the flange to yield until there is full contact between spoke and flange. When spokes are properly tensioned, the aluminum flange material on which they bear is usually under enough stress to conform to the spokes.

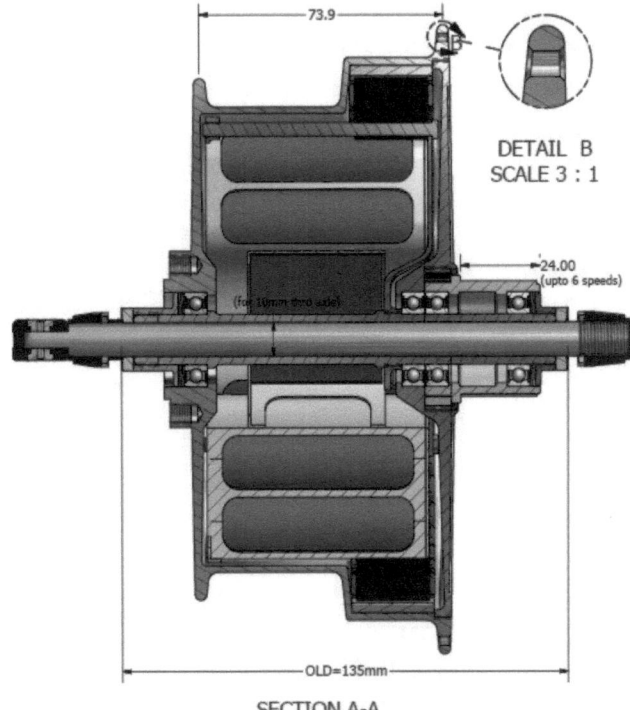

Figure 30: The cross-sectional of the E-Wheel™

To give better spoke support and to allow easier spoke insertion, the edges of spoke holes are usually beveled *(refer the detail B in the figure above)*. Some aluminum alloy hubs are made with trumpet-shaped holes to match the curvature of the spoke elbow. However, a plain hole in which the spoke forms its own contour gives better support than a preformed radius that invariably does not match the spoke elbow.

The design of the hub based on *DT Swiss® 545d 26"* rim with single eyeleted, feature a wide, reinforced profile; *DT alpine III®* 2.0 – 1.8 – 2.34mm (gauges 14|15|13) spokes and *DT standard brass nipples*. [30]

The wheel supports static and dynamic loads that have radial, lateral and torsional components. Both static and dynamic loads cause stresses in the wheel. Static loads, such as spoke tension or tire inflation pressure, remain constant or change infrequently. Dynamic loads change continually. Despite of being constant, the rider's weight is a dynamic load because the wheel's rotation causes it to produce changing forces within the wheel. The hierarchy ralationship of bicycle wheel loads are shown as the following:

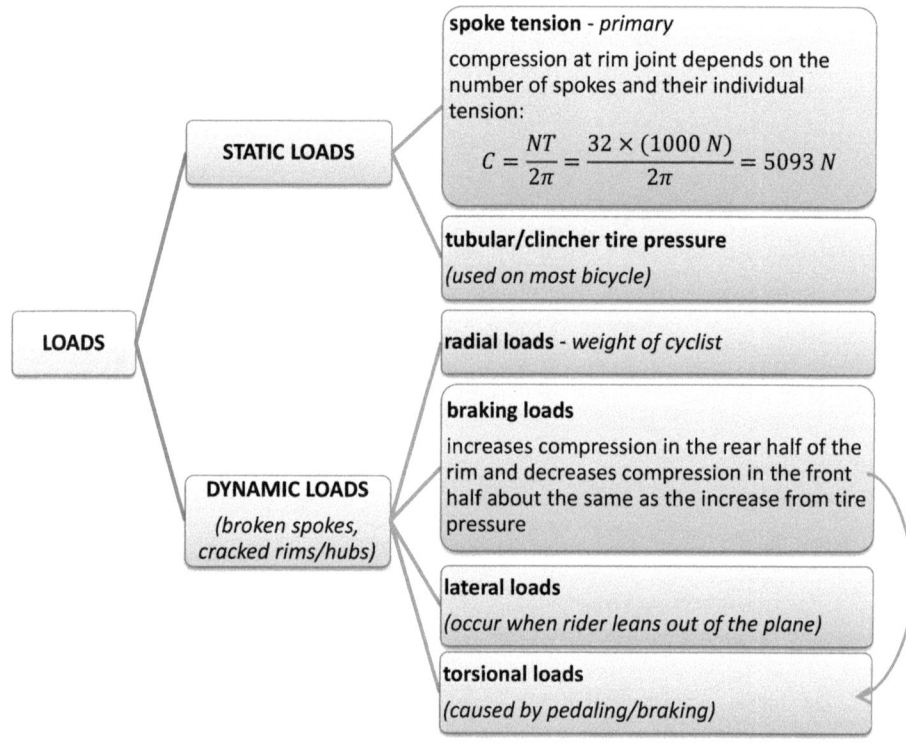

Figure 31: Loading calculation hierarchical relationships

Various structure formulas were used to determine the values, which are reported. Some of these equations are shown below:

Torsional stiffness of the hub shaft [34]

$$K_t = \frac{JG}{L}, (Nm/deg)$$

where:

- Shear modulus of the hub shaft material staintless steel, in this case: $G = 77.2 \ (GPa)$

- Moment of inertia, J:

$$J = \frac{\pi}{32}(D^4 - d^4) = \frac{\pi}{32}[(0.015\ m)^4 - (0.01\ m)^4] = 3.99 \times 10^{-9}\ (m^4)$$

- Length of shaft: $L = 128\ (mm)$

Thus, torsional stiffness is:

$$K_t = \frac{(3.99 \times 10^{-9}\ m^4) \times (77.2\ GPa)}{128\ (mm)} \approx 2406.5\ (Nm/rad)$$

$$\rightarrow K_t = \frac{\pi \times 2406.5\ (Nm/rad)}{180} \approx 42\ (Nm/deg)$$

Spoke tension and deflection can be measured by a **spoke tensiometer**. Although these methods work well enough to verify theory, they are cumbersome and not sufficiently repeatable to analyze wheel deflections precisely. For a precise analysis of deflections, a mathematical model was developed by using the finite element method (FEM) for structural analysis.

The maximum spoke tension of 545d rim is $1200N$, [30] using the pre-set at $1000N$ is suitable.

Building wheels with the correct length spokes makes things so much easier. Building with the wrong length spokes can often mean that the wheel will not build and the makers only realize this half way through the build sequence (or later).

Spoke length can be easily found down by appling *the spoke length formula:* [35]

$$spoke\ length = \sqrt{R^2 + H^2 + F^2 - 2RH \cos\left(\frac{720}{h}X\right)} - \frac{\phi}{2}$$

where:

- R: Rim radius to spoke ends
- H: Hub radius to spoke holes
- F: Flange offset
- X: Lacing pattern ($X = 3$ in this case)

- h: Holes in rim

- ϕ: Diameter of spoke hole in hub

In addition, the hub has high-lows flange so there are two different length of spokes. As with the rims there are only a few key dimensions on the hub that required for spoke length calculations, as shown in the below table:

	Left side of the hub	Right side of the hub
Flange diameter (mm)	147.0	179.0
Offset (from) (mm)	44.0	30.0
Spoke length (mm)	256.1	252.5

Table 13: Key dimensions of the hub

Very complex load conditions [21] for sitting – pedaling and braking must be applied. To estimate **sitting – pedaling loads** – asymmetrical load conditions due to chain effect – changeover either 481N and 489N for each side of pedals.

How to estimate **braking loads**? Braking with a disc brake causes a small but significant radial load that affects spoke tension. Normally, under hard braking, the disc brake cause the rim with a force of up to 800N by pushing rearward with 400N force and pulling on the front half of the rim equally. This increases compression in the rear half of the rim and decreases compression in the front half about the same as the increase from tire pressure.

The design of E-Wheel™ is only significantly compatible with 180mm or 203mm disc brake rotor, smaller rotor, 160mm is not encourage to be used.

Using disc brake in this design, thus the rotor diameter is one of the contributing factor to braking power or braking loads. Large diameter rotors provide more leverage for the brake system to oppose rotation of the wheel by placing the braking force further from the axle. Therefore, the larger the rotor is, the more powerful the brake is. Larger rotors are unavoidably heavier than their smaller counterparts, so riders may wish to run smaller rotors for overall weight savings. However, this could lead to a brake system that isn't adequately powerful for heavier or more aggressive riders.

3.6 Axle

The shaft (axle) carrying stator of the BLDC Motor, hub, and so on must be supported by bearings. Two ball bearings can provide sufficient radial support to limit shaft bending and deflection to acceptable values, this is highly desirable and simplifies manufacturing.

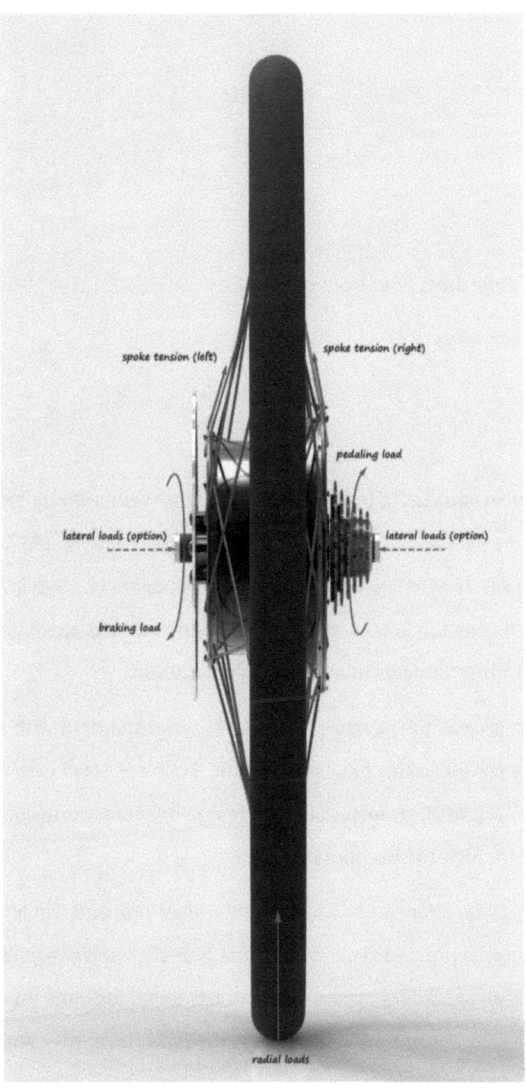

Figure 32: Forces applied to the E-Wheel™

In order to compatible with thru-axle or quick release mechanism, the shaft must be stationary hollow one and requests durability. It is unnecessary to evaluate the stresses in a shaft at every point, a few potentially critical locations will suffice. Perform free body diagram analysis to get reaction forces at the bearings in the hollow axle [19]

As we both know, the stress in the bending beam is then just a function of the bending moment: [20]

$$\sigma = \frac{Mc}{I}$$

where:

- M: Bending moment
- c: Distance of farthest fiber from neutral axis
- I: Moment of Inertia

3.7 Bearing

When the first bicycle appears to have been invented by Leonardo da Vinci *[James McGurn, An Illustrated History of Cycling, John Murray Publishers, London, 1987. T.A Harris, Rolling Bearing Analysis, 1991 John Wiley & Sons]*, who also appears to have invented the first ball bearing. However, it took the 1800's industrial revolution and its capability to precisely work metal on a large commercial scale to make bikes ubiquitous.

As the same with this case, the design of E-Wheel™ come standard with the simplest, hence lowest cost, 61902 (or 6902-2rs) fully sealed 440C Stainless Steel cartridge bearings, from **SKF Corporation**, using 440C Stainless Steel to resist moisture, corrosion and abrasion, and for their suitability for high speeds operation.

To work effectively, interference fits between the outer ring and the hub must be applied, which mean the outer is pressed securely into the hub shell and the inner raceway must be held securely between shoulders on the axle and tightened locknuts. At the same time, the inner and outer raceways must be aligned so that the balls run in the center of the raceways.

Interference fits in general only provide sufficient resistance to axial movement of a bearing on its seating when no axial forces are to be transmitted and the only requirement is that lateral movement of the ring should be prevented. Positive axial location or locking is necessary in all other cases. To prevent axial movement in either direction of a locating bearing it must be located at both sides.

4 BLDC Motor Controller Design

4.1 Design Strategy and Goals

This chapter presents a apply-oriented solution for control of brushless DC motor for E-Wheel™ using microcontrollers and integrated circuit from a leading brand, Texas Instruments Incorporated, that enable the cost-effective design of intelligent controllers for three-phase motors by reducing the system components and increasing efficiency.

The economic constraints and new standards legislated by governments place increasingly stringent requirements on electrical systems. New generations of equipment must have higher performance parameters such as better efficiency and reduce electromagnetic interference. System flexibility must be high to facilitate market modifications and to reduce development time. All these improvements must be achieved while, at the same time, decreasing system cost.

For the first experience as a Control Designer, this application report only covers the following:

- A theoretical background on field oriented motor control principle
- Controller structures

Further application must be represented by the full-of-skill control designers based on Electric/Mechatronic.

4.2 Controller Design

The key to effective torque and speed control of a BLDC motor is based on relatively simple torque and back EMF equations, which are similar to those of the DC motor. The back EMF magnitude can be written as:

$$E = 2NIrBw$$

And the torque term as:

$$T = \left(\frac{1}{2}i^2\frac{dL}{d\theta}\right) - \left(\frac{1}{2}B^2\frac{dR}{d\theta}\right) + \left(\frac{4N}{\pi}Brl\pi i\right)$$

where N is the number of winding turns per phase, l is the length of the rotor, r is the internal radius of the rotor, B is the rotor magnet flux density, w is the motor's angular velocity, i is the phase current, L is the phase inductance, θ is the rotor position, R is the phase resistance.

The first two terms in the torque expression are parasitic reluctance torque components. The third term produces mutual torque, which is the torque production mechanism used in the case of BLDC motors. To sum up, the back EMF and the torque production is directly proportional to the motor speed and to the phase current respectively. These factors lead to the BLDC motor speed control schemes as shown in below figure:

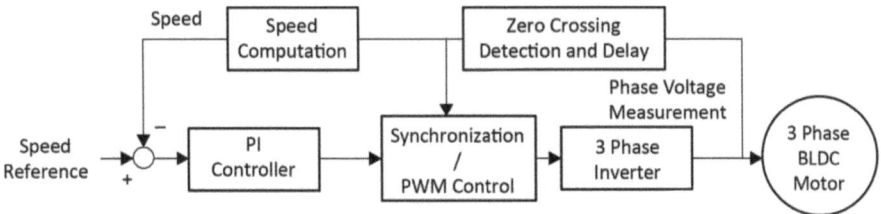

Figure 33: Speed and current control loop configurations for a BLDC motor

The BLDC motor is characterized by a two phase ON operation to control the inverter. In this control scheme, torque production follows the principle that current should flow in only two of the three phases at a time and that there should be no torque production in the region of the back EMF zero crossings. *Figure 34* describes the electrical wave forms in the BLDC motor in the two phases ON operation.

This control structure has several advantages:

- Only one current at a time needs to be controlled
- Only one current sensor is necessary (or none for speed loop only, as detailed in the next sections)
- The positioning of the current sensor allows the use of low cost sensors as a shunt

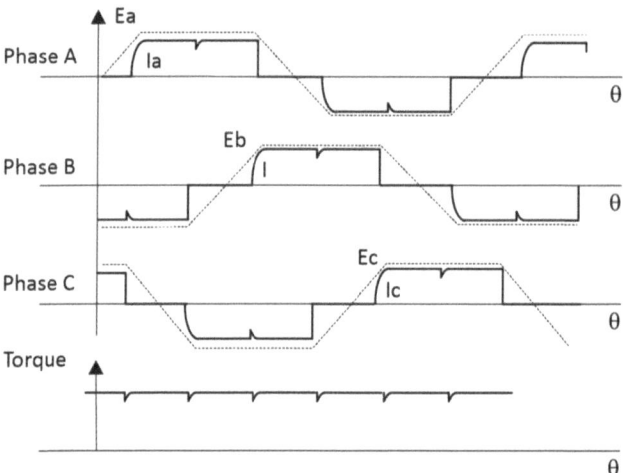

Figure 34: Electrical waveforms in the two phase ON operation and torque ripple

The principle of the BLDC motor is, at all times, to energize the phase pair, which can produce the highest torque. To optimize this effect, the back EMF shape is trapezoidal (as in the previous section). The combination of a DC current with a trapezoidal back EMF makes it theoretically possible to produce a constant torque. In practice, the current cannot be established instantaneously in a motor phase; as a consequence the torque ripple is present at each 60° phase commutation.

Limited of placement space is one of the most important constraint for motor controller design, thus SMD (Surface Mount Devices) must be used instead of thru-hole devices. These components are usually smaller than its through-hole counterpart because it has either smaller leads or no leads at all.

4.2.1 Three phase inverter MOSFETs

The vast majority of motor controllers use only N-channel power MOSFETs, and surface mount with leads in this case (D^{2}pak package). Normally, MOSFETs have three accessible terminals (or pins): a gate, a drain and a source. The presence of an anti-parallel "body-diode" is indicated in the symbol; this diode is part of the MOSFET and will *always* conduct current from source to drain if the voltage of the source is higher than that of the drain. As with any power transistor, the function of the MOSFET is to convert a low-power signal into an amplified high-power output.

This can be seen in typical set of power MOSFET I-V curve, such as in *Figure 35*. This curve is set for the International Rectifier MOSFET IRFS3006PbF [22].

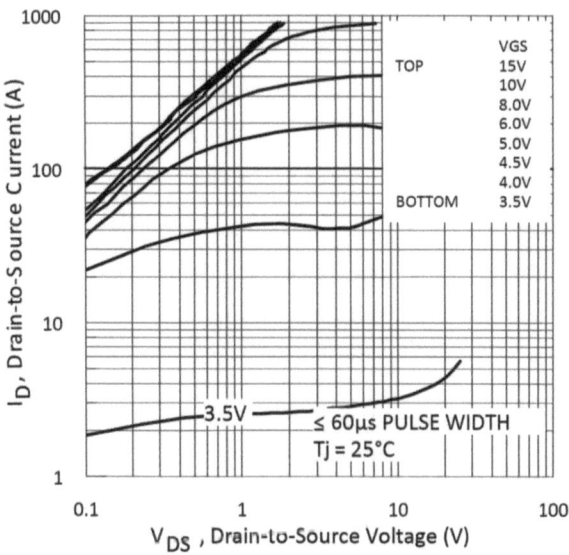

Figure 35: A typical set of power MOSFET I-V curve

The below table lists some other relevant specifications for this MOSFET. The first thing to check is that the MOSFET voltage rating is a good deal higher than the system voltage, which in this case it is 36V from battery source. With an on-state resistance of $0.025m\Omega$, the power dissipated in the MOSFET at $20A$ would be:

$$P_{dissipated} = I^2 R_{DS(on)} = (20A)^2 (0.0025\Omega) = 1W$$

Specification	Value
V_{DSS}	60V
$R_{DS(on)}\ max$	$2.5m\Omega$
$R_{\theta JA}$	40°C/W
$R_{\theta JC}$	0.4°C/W
$T_J\ max$	175°C

Table 14: Some specifications for the IR MOSFET IRFS3006PbF

MOSFET datasheets usually give multiple thermal resistance values. One, here labeled $R_{\theta JA}$, may denote the junction-to-ambient thermal resistance, if there is no heat sinking or forced convection. In this case, with $1W$ dissipation, the MOSFET can be operated with free-air convection only. This and the heating of bus capacitors represent the two biggest limitations on continuous current capacity of a motor controller.

The BLDC motor control consists of generating DC currents in the motor phases. This control is subdivided into two independent operations: stator and rotor flux synchronization and control of the current value. Both operations are realized through the three phase inverter depicted in *Figure 36*.

The flux synchronization is derived from the position information coming from sensors, or from sensorless techniques. From the position, the controller determines the appropriate MOSFETs (M1 to M6) that must be driven. The regulation of the current to a fixed 60° reference can be realized in either of the two different modes:

- The Pulse Width Modulation (PWM) Mode: The supply voltage is chopped at a fixed frequency with a duty cycle depending on the current error. Therefore, both the current and the rate of change of current can be controlled.

- The Hysteresis Mode: The power transistors are switched off and on according to whether the current is greater or less than a reference current.

This design uses the pulse width modulation mode for control the BLDC motor according to *Figure 37*.

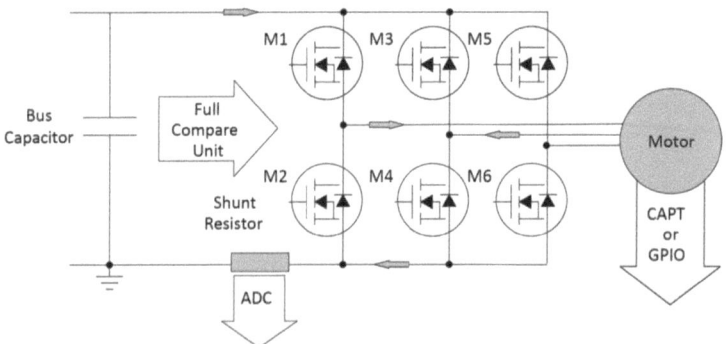

Figure 36: The bus capacitor is placed across the DC lines, adjacent to the inverter MOSFET (Three phase inverter)

4.2.2 Bus Capacitor

Bus capacitors are the most important passive component in a motor controller. They source and sink high instantaneous currents to and from the inverter, in affect shielding the rest of the controller from the high frequency switching transients. They also accommodate for battery (or power supply) cable inductance, absorbing energy that would otherwise create damaging voltage spikes when current is suddenly switched off. *Figure 36* shows the placement of the bus capacitor; it is crucial that the capacitor be physically located as close as possible to the inverter MOSFETs, so that the resistance and inductance of the line between the capacitor and the MOSFETs is very low.

The bus capacitor is typically an electrolytic capacitor or a group of electrolytic capacitors in parallel. In high-current controllers, it can occupy as much or more volume than the MOSFETs themselves. It can also dissipate as much or more heat, meaning it contributes significantly to the overall efficiency of the controller. Sizing the bus capacitor is critical to a motor controller design. There are two main considerations for sizing: voltage ripple and heating.

Voltage ripple is easy to predict by making the worst-case assumption that the bus capacitor supplies all of the switching current at the PWM frequency. In this case, the power supply provides only a steady DC current. When the PWM is off, the bus capacitor is discharged, sourcing current (in addition to the power supply current) to the MOSFETs. *Table 15* lists the currents present in these two states:

	Power Supply Current	MOSFET Current	Capacitor Current
PWM Off (1-D)	$(D)I_{motor}$	0	$(D)I_{motor}$
PWM On (D)	$(D)I_{motor}$	I_{motor}	$-(1-D)I_{motor}$

Table 15: Current flow during the PWM on and off times, assuming the power supply provides only a DC average current (high power supply inductance and/or high frequency)

The duty cycle, D, is the fraction of time when PWM is on and $(1-D)$ is off. By taking the time-weighted sum of the currents in Table 15, the average current in from the power supply over one switching cycle is the same as the current out of the MOSFET. Additionally, the average capacitor current is zero, a necessary condition for steady-state.

The voltage ripple depends on the current, the duty cycle, and the switching frequency. Starting from the constitutive equation for a capacitor, it can be calculated as follows:

$$\frac{dV}{dt} = \frac{I}{C}$$

$$\Delta V = \frac{I}{C}\Delta t = \frac{DI_{motor}}{C}\frac{(1-D)}{f_{sw}}$$

This is the calculation for charging. Discharging will yield the same equation, but with the negative change in voltage. (So that over one switching cycle in steady-state, the capacitor returns to the same voltage.). From this, it is easy to see that the maximum ripple voltage occurs at 50% duty cycle, where $D(1-D) = 0.25$.

The capacitor for inverter in this controller design is $330\mu F, 50V$ aluminum electrolytic SMT capacitor, manufactured by Panasonic Electronic Components. [25]

The other important design consideration in sizing the bus capacitor is heat generation. Calculating the capacitor heat loss is impossible without knowing the ripple current more exactly. The worst-case assumption, though, is the same as in the voltage ripple calculation: the bus capacitor supplies the entire AC component of current, at 50% duty cycle. One way to very roughly estimate capacitor heating is using the "loss tangent", which is defined as follows:

$$tan\delta = \omega CR$$

The value $tan\delta$ is published in capacitor datasheets, and is assumed to be constant for all frequency ω. C is the capacitance and R is an equivalent series resistance. The power dissipation in the capacitor is definitely high enough to significantly influence the overall controller efficiency. It may also be high enough to heat the capacitor above its maximum temperature.

4.2.3 Gate Driver

The DRV8x family of integrated motor drivers from Texas Instruments enables manufacturers to quickly and easily spin their motors. Integrated drivers provide higher performance and better protection within a smaller board footprint versus traditional discrete solutions. Furthermore, integrated drivers are simpler and faster to design because they do not require discrete drive-stage design experience.

The design of E-Wheel™ controller based on the simplest **DRV8301**, [23] gate driver IC for three phase motor drive applications. It provides three haft bridge drivers, each capable of driving to **N-type MOSFETs**, one for the high-side and one for the low-side. The DRV8301 includes two current shunt amplifiers for accurate current measurement. The current amplifiers support bi-directional current sensing and provide an adjustable output offset up to $3V$. It also has an integrated switching mode buck converter with adjustable output and switching frequency to support MCU or additional system power needs. The buck is capable to drive up to $1.5A$ load.

Description	Supply Voltage (V)	I_{OUT} Cont. (A)	I_{OUT} Peak (A)	Control Interface	Drives Solenoid	Op Temp (°C)	Price[*]
Pre-driver with 1.5-A step-down voltage regulator and dual current-sense amps (SPI control)	6 to 60	Ext FETs	Ext FETs	PWM	No	-40 to 125	2.50

Table 16: Some Specifications of DRV8301

The SPI interface provides detailed fault reporting and flexible parameter settings such as gain options for current shunt amplifier, slew rate control of gate driver, etc.

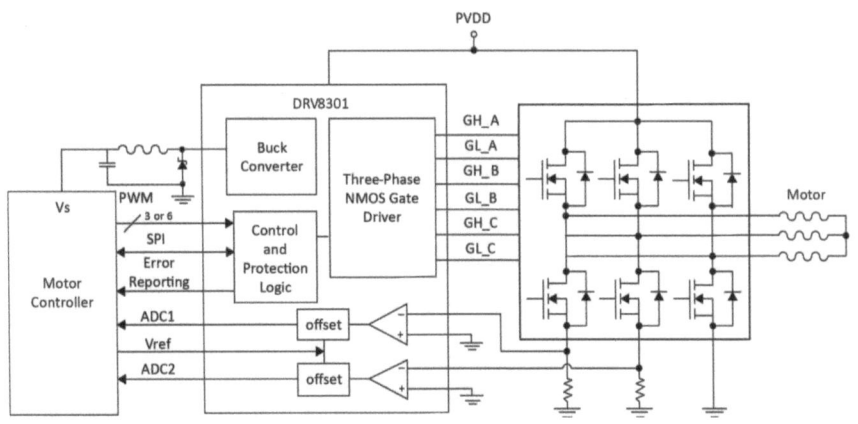

Figure 37: DRV8301 Simplified Application Schematic

---

[*] Suggested resale price in U.S. dollars in quantities of 1,000.

The DRV8301 are defined **"Green"** to compatible with Restriction of Hazardous Substances (RoHS) for electronic and electric devices, which mean free of Lead (Pb), Bromine (Br) and Antimony (Sb) based flame retardants (Br or Sb do not exceed 0.1% by weight in homogeneous material).

4.2.4 Motor Controller

Using a microcontroller (MCU)-based three-phase BLDC pre-driver with a companion IC to perform power management demonstrates a much simpler design. MCU suppliers (such as Texas Instruments or Microchip, etc.) offer dedicated peripherals for motor-control applications to optimize the system performance. There are also solutions that integrate the controller and pre-driver into one package, resulting in a smaller-footprint design. One disadvantage of the fully-integrated solution is that both the I/O and computing power are fixed, limiting the design's flexibility.

Today, the amount of PCB space consumed and the thickness of the enclosure are concerns for many three-phase BLDC applications. An integrated three-phase BLDC motor controller reduces system complexity and helps to meet the minimum PCB space requirements. However, the maximum motor drive strength is limited by its power dissipation, package selection and the silicon design.

Typically microcontroller includes detection, A/D conversion, and output comparison components. For a low-end system, the ultra-low power microcontroller is available while the Texas Instruments C2000 digital signal processor can be used for a complex system with more features. The F2802x Piccolo™ family of microcontrollers provides the power of the C28x™ core coupled with highly integrated control peripherals in low pin-count devices. This family is code-compatible with previous C28x based code, as well as providing a high level of analog integration.

The 2802x (C28x) family is a member of the TMS320C2000™ microcontroller (MCU) platform. The C28x-based controllers have the same 32-bit fixed-point architecture as existing C28x MCUs. It is a very efficient C/C++ engine, enabling users to develop not only their system control software in a high-level language, but also enabling development of math algorithms using C/C++. The device is as efficient at MCU math tasks as it is at system control tasks that typically are handled by microcontroller devices. This efficiency removes the need for a second processor in many systems. The 32 x 32-bit MAC 64-bit processing capabilities enable the controller to handle higher numerical resolution problems efficiently.

CPU Speed (MHz)	Flash (KB)	RAM (KB)	PWM channel	High-resolution PWM channel	Event captures	Timers	12-bit ADC channel	Comparators	ADC conversion time	ADC Speed (kSPS)	I²C	UART/SCI	SPI	Core supply (V)	GPIO pins	Package pin counts	Temperature range (°C)	1 kU pricing (U.S. $)
60	64	12	9	4	1	9	13	2	217	4600	1	1	1	3.3	22	38 TSSOP, 48 QFP	-40 to 125	3.55

Table 17: TMS320F28027 Piccolo™ microcontroller simplified specifications [24]

This design based on the TMS320F28027 Piccolo™ microcontroller from TI. An internal voltage regulator allows for single-rail operation. Enhancements have been made to the high-resolution PWM module to allow for dual-edge control (frequency modulation). Analog comparators with internal 10-bit references have been added and can be routed directly to control the PWM outputs. The ADC converts from 0 to $3.3V$ fixed full-scale range and supports ratio-metric V_{REFHI}/V_{REFLO} references. The ADC interface has been optimized for low overhead and latency.

4.2.5 Position and speed sensing

The Hall effect has been known for over one hundred years, but has only been put to noticeable use in the last three decades. Today, Hall effect devices are included in many products, ranging from computers to sewing machines, automobiles to aircraft, and machine tools to medical equipment.

The commutation of BLDC motor is controlled electronically. To rotate the BLDC motor, the stator windings should be energized in a sequence. It is important to know the rotor position in order to understand which winding will be energized following the energizing sequence. Rotor position is sensed using Hall effect sensors embedded into the stator. Most BLDC motors have three Hall effect sensors embedded into the stator on the non-driving end of the motor.

Whenever the rotor magnetic poles pass near the Hall sensors, they give the high or low signal, indicating the **N** or **S** pole is approaching the sensors. Based on the combination of three Hall effect sensor signals, the exact sequence of commutation can be determined.

Rotor position is reported by three built-in Hall sensors which deliver six different signal combinations per commutation sequence. The three phases are powered in six different conducting phases in line with this sensor information. The current and voltage curves are block-shaped. The switching position of every electronic commutation lies symmetrically around the respective torque maximum.

Figure 37 illustrates how this electronic commutation can be performed by three digital output Hall Effect sensors and microcontroller. Permanent magnet materials mounted on the rotor shaft operate the sensors. The sensors sense the angular position of the shaft and feed this information to a logic circuit. The logic circuit encodes this information and controls switches in a drive circuit. Appropriate energized windings, as determined by the rotor position, are magnetic field generated by the windings rotates in relation to the shaft position. This reacts with the field of the rotor's permanent magnets and develops the required torque.

The below schematic represents overall structure of Hall Effect sensor using with the microcontroller to control the BLDC motor.

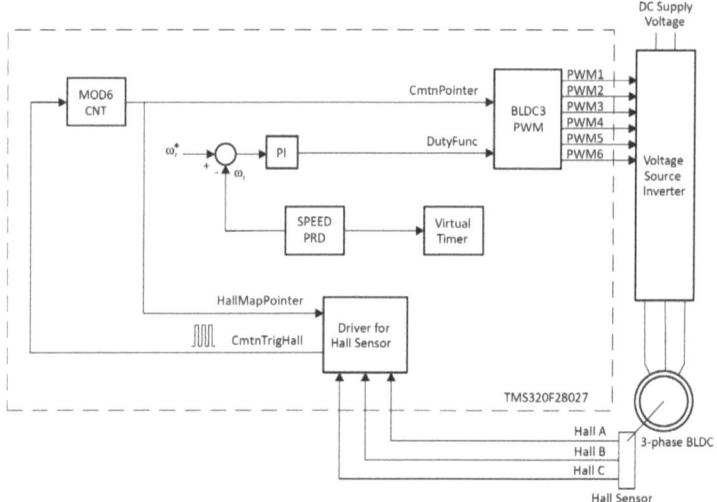

Figure 38: Overall block diagram of Hall sensors and BLDC motor microcontroller

4.2.6 Power Supply

An important interface between power and signal sections of a controller is at the power supplies. The power for microcontrollers, inverters, and all signal-level devices is derived from the primary battery through several stages of the power conversion. There are three basically level in the system, all are oriented from a $+36V$ Lithium-ion battery (as in the previous section):

$+36V$ from the Lithium-ion battery.

$+15V$ is used for powering the Gate driver.

$+3.3V$ specifies the power supply of Gate driver IC, microcontroller and other devices.

The first power supply is an efficient switching regulator, which converts battery voltage ($+36V$) to $+15V$. And the output of the $+15V$ supply is also used to create the signal-level power supply ($+3.3V$) by using step-down linear regulator. Linear regulator produce a clean output (no switching), which is good for signal electronics. However, they dissipate a lot more heat than switching regulators.

Large value electrolytic capacitors provide bulk charge for motor transitions as well as conducted emissions filtering.

But, at first, using *Mean Well NES-350-36 AC-DC* power supply as a replacement for Lithium-ion battery for testing system.

4.2.7 Battery Management

Most battery packs include some type of protection to safeguard battery and equipment, should a malfunction occur. The most basic protection is a fuse that opens if excessively high current is drawn. A more complex protection circuit is found in intrinsically safe batteries. These batteries are mandated for two-way radios, gas detectors and other electronic instruments. The protection circuit prevents excessive current, which could lead to high heat and electric spark.

Figure 39 is a block diagram of circuitry in a typical Lithium-ion battery pack. It shows an example of a safety protection circuit for the Lithium-ion cells and a gas gauge (capacity measuring device). The safety circuitry includes a Lithium-ion protector that controls back-to-back FET switches. These switches can be opened to protect the pack against fault conditions such as overvoltage, under voltage, and overcurrent. The diagram also includes a temperature sensitive three-terminal fuse that will open due to prolonged overcurrent or over temperature, or it can be forced to open by redundant protection circuitry in case there is a fault where the primary protection circuitry fails to respond. Opening this fuse is a last resort, as it will render the pack permanently disabled.

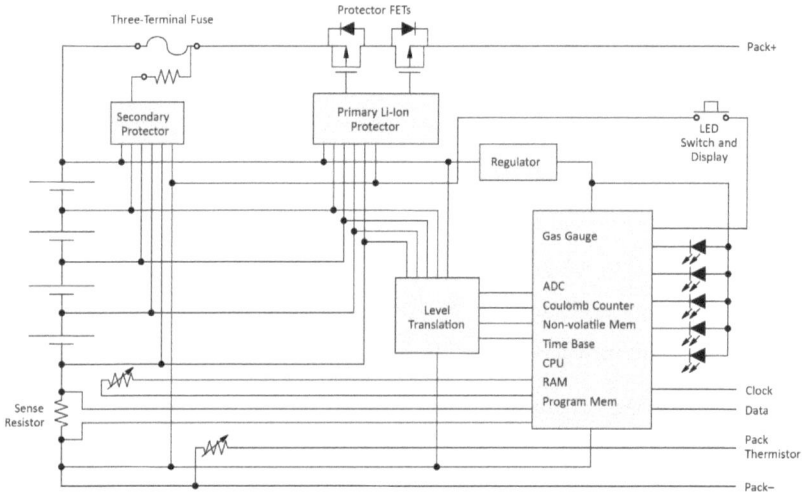

Figure 39: Block diagram of circuitry in a typical Lithium-ion battery pack

The gas-gauge circuitry measures the charge and discharge current by measuring the voltage across a low-value sense resistor with low-offset measurement circuitry. The current measurement is integrated to determine the change in coulometric capacity. In addition, the gauge measures temperature and voltage, evaluates gas-gauging algorithms to determine the available capacity in the battery, and computes time-to-empty and other values required by the host. The available capacity as well as other measurements and computational results are communicated to the host over a serial communication line. A visual indication of available capacity can be displayed by the LEDs when activated by a push-button switch.

Conclusions

Among the development trends, people love the new, this design can encourage an equally powerful attraction: *love of classic*. Classic bikes from 50 years ago look the same today, and they are still in style with E-Wheel™. The emphasis is on *"style"* beside on the competitive advantage of unprecedented design and sweat-free lack of effort.

The E-Wheel™ presented in this application report is a new kind of pedelec and was designed to fix into any conventional bikes. With the increasing demand for eco-friendly mobility transportation, E-Wheel™ offer an improved quality of pedaling to everybody, especially since E-Wheel™ has had some initial successes, e.g. design awards and feedback from entire the world. And the author hopes that E-Wheel™ will be put on the road and received comment from communities in the near future.

This project has been an extremely rewarding pursuit for the author and no written report can fully capture the nature of learning experience afforded by pursuing design the E-Wheel™. The electronics are at the center of attention in this concept which aim to create new kind of mobility. For this reason, it is the author's hope that others will find parts of this report useful for their own works.

References

[1] *Finite Element Method Magnetics tutorial* http://www.femm.info/wiki/HomePage

[2] *JMAG Self-learning*, JMAG International https://www.jmag-international.com/index.html

[3] Phuoc Nguyen, *"End-of-trial Report to JSOL Corporation"*, internal report to New System Vietnam Ltd. and JSOL Corporation Japan, September, 2014.

[4] Shane W. Colton, *"Design and Prototyping Methods for Brushless Motors and Motor Control"*, graduate thesis of Degree of Master of Science in Mechanical Engineering, Massachusetts Institute of Technology, June, 2010.

[5] Dr. Duane Hanselman, *"Brushless Permanent Magnet Motor Design"*, 2nd edition, Magna Physics Publishing, 2006.

[6] C. Pompermaier, L. Sjöberg and G. Nord, *"Design and optimization of a Permanent Magnet Transverse Flux Machine"*, in International Conference on Electrical Machines (ICEM), 2012.

[7] Thomas Lewis, *"Go Pedelec Handbook"*, Go Pedelec Project Coordinator & Country Manager Austria *energieautark consulting gmbh*, Hauptstraße 27/3 A-1140 Wien, Austria. www.energieautark.at, www.gopedelec.eu

[8] Annick Roetynck, *"PRESTO Cycling Policy Guide Electric Bicycles"*, ETRA Secretary General, Intelligent Energy Europe, Belgium, February 2010.

[9] *"Panasonic Lithium Ion Batteries Technical Handbook"*, Panasonic Corporation, June 2007.

[10] *"Copenhagen Wheel"*, MIT Senseable City Lab, Cambridge, USA. https://www.superpedestrian.com/

[11] *"FlyKly Smart Wheel"* http://www.flykly.com/

[12] *"Zehus BIKE+"*, Zehus s.r.l Italy, http://www.zehus.it/

[13] *"Electron Wheel"* http://www.electronwheel.com/

[14] *"Lexus Design Award 2014"*, organized by Lexus International, co-hosted by Designboom and DESIGN ASSOCIATION NPO, http://www.lexus-int.com/design/lda.html.

[15] Danielle Demetriou, *"A new generation in design – How the Lexus Design Award is shaping the future of design"*, Lexus International, http://www.lexus-int.com/design/lexus-design-award/

[16] *"ASEANpreneurs Autodesk Design Challenge"*, organized by Autodesk Inc. USA and NUS Entrepreneurship Society, http://aseanpreneurs.org/autodesk_ap_design/

[17] *"Polyacetal Catalog"*, Quadrant Engineering Plastic Product, Quadrant Group.

[18] *"Design for Product Lifetime – Design for Durability"*, Autodesk Sustainability Workshop, Autodesk Inc. USA.

[19] *"Machine Design Databook"*, Digital Engineering Library, Mc-Graw Hill, 2014.

[20] Richard G. Budynas, J. Keith Nisbett, *"Shigley's Mechanical Engineering Design"*, 9th edition, Mc-Graw Hill, 2011.

[21] L.Maestrelli, *"Bicycle frame optimization by means of an advanced Gradient Method Algorithm"*, 2nd edition, European HTC Strasbourg, September 30th – October 1st , 2008.

[22] International Rectifier IRFS3006PbF MOSFET datasheet, http://www.irf.com/product-info/datasheets/data/irfs3006pbf.pdf

[23] Texas Instruments Three-Phase Gate Driver DRV3801 datasheet, http://www.ti.com/lit/ds/symlink/drv8301.pdf

[24] Texas Instruments Piccolo™ microcontroller TMS320F28027 datasheet, http://www.ti.com/lit/ds/symlink/tms320f28027.pdf

[25] Panasonic Surface Mount Type Aluminum Electrolytic Capacitors Catalog, Panasonic Corporation, 2014.

[26] NdFeB/Neodymium Iron Boron Magnets Datasheet, E-Magnets UK.

[27] Padmaraja Yedamale, *"Brushless DC (BLDC) Motor Fundamentals"*, Microchip Technology Inc., 2003.

[28] Ward Brown, *"Brushless DC Motor Control Made Easy"*, Microchip Technology Inc., 2002.

[29] *"Machining Instructions for Engineering Plastics"*, Gilbert Curry Industrial Plastics Co. Ltd. and QUADRANT Engineering Plastic Products.

[30] DT Swiss website http://www.dtswiss.com/Components

[31] Christo Skylar, *"Winding Scheme Calculator"*, http://www.bavaria-direct.co.za/scheme/calculator/

[32] *"Selection of Electrical Steels for Magnetic Cores"*, AK Steel Corporation, 2007.

[33] John Barnett, *"Barnett's Bicycle Repair Manual"*, 9th edition, Barnett Bicycle Institute, 2014.

[34] Jobst Brandt, *"The Bicycle Wheel"*, 3rd Edition, page 129, equation 5.

[35] Roger Musson, *"A practical Guide to Wheel Building"*, 3rd Edition, page 79.

[36] Nanette Wong, *"Lexus Design Award 2014 Winners"*, Design Milk Magazine, http://design-milk.com/lexus-design-award-2014-winners-part-1/

[37] George D. Gopen, Judith A. Swan, *"The Science of Scientific Writing"*, American Scientist, Volume 78, 550-558, Nov-Dec 1990.

Appendices

#	Parts Name	Quantity	Order No	Manufacturer	Distributor/Dealer	Price	Unit	REV
1	Power Supply	1	NES-350-36	Meanwell	Hiep Luc Ltd	1200 000	VND	distributor store
2	InstaSPIN™-FOC enabled C2000 Piccolo LaunchPad	1	LAUNCHXL-F28027F	Texas Instruments	-	17	USD	shop online
3	25x10x3mm block of 5 magnets	9	EP647	e-Magnets UK	-	7.65	GBP	shop online
	1"x3/8"x1/8" thick Nickel Plated magnet	44	BX062	K&J Magnetics Inc	-	1.65	USD	shop online
4	Delrin Acetal 150mm dia x 80mm long rod	1	Erta-cetal H	Quadrant Plastics	theplasticshop.uk	26.5	GBP	shop online
5	Copper magnet wire coil AWQ-22 8oz 251ft 155C	1	MW0038	TEMCo Industrial Power Supply	-	8.21	USD	shop online
6	Lamination steel	60	1405147	PROTO Laminations Inc	-	10.49	USD	quote
		47	7713	Polaris Laser Laminations, LLC	-	7.65	USD	quote
7	Lithium-ion battery cell	30	NCR-18650A	Panasonic	-	1500 000	VND	distributor store
8	Ball bearing	2	W61902-2RS1	SKF	-	-	-	distributor store
9	Rim	1	545d	DT Swiss	-	25	USD	distributor store
10	Spoke	36	DT alpine®	DT Swiss	-	-	-	-
11	Nipple	36	DT standard brass	DT Swiss	-	-	-	-
12	Shimano Uniglide® splines freehub	1		Zipp Speed Weaponry	-	100	USD	shop online

Table 18: Estimating Financial Expenditures, Suppliers and Part/Service Lists for Prototyping

By carefully selecting the manufacturers or suppliers of each part of the design, the author will make sure to fit-on-budget, and also contribute to a best quality and sustainable product.